RADAR WATCHKEEPING

Captain W. D. Moss, Master Mariner

with a foreword by
Captain J. H. Quick, C.B.E., F.I.N.
*Professional Officer and Chief Nautical Surveyor,
Ministry of Transport*

THE MARITIME PRESS LIMITED
(MEMBER COMPANY OF THE GEORGE PHILIP GROUP)
12–14 Long Acre · London WC2

First edition © 1965

Second edition
Revised and enlarged © 1973

The Maritime Press Limited

All rights reserved. No part of this publication
may be reproduced, stored in a retrieval system,
or transmitted in any form or by any means,
electronic, mechanical, photocopying, recording
or otherwise, without the permission of the
publishers.

ISBN 0 540 00964 4

*Printed in Great Britain by
Ebenezer Baylis and Son Limited
The Trinity Press, Worcester, and London*

Foreword

Despite what has been said and published in the past about the proper and effective use of Radar in avoiding collisions we still have far too many collisions between ships fitted with Radar. When these casualties are investigated it seems that in almost every case disaster has occurred through human failure either to appreciate what the Radar screen is conveying or to act in a prudent and responsible manner on the information which the screen provides.

In this book Captain Moss has attempted to produce a really down-to-earth approach to this problem, leaving aside the more theoretical niceties which may not be everybody's meat.

On the eve of my departure from the position which I have been privileged to occupy for a number of years I give this book my blessing, and hope that apart from being readily readable by all mariners and a valuable contribution to the understanding of Radar, it may inspire those who do not behave as they should to think afresh.

There is no doubt that Radar has enabled the competent seaman to prosecute his voyage in safety and with expedition; but as I see it there remain other categories (fortunately in a minority)—those who have had training but consider it so much "eye wash" and those who see no need for any instruction at all. May they be converted.

JAS. H. QUICK

Contents

INTRODUCTION

1	Why collisions occur despite radar	7
2	The radar set	12
3	Miscellaneous effects and errors	21
4	Radar information and its application	31
5	Radar plotting	41
6	Proceeding in reduced visibility	70
7	Coastal navigation	85
8	Radar on small craft	98
9	Clear weather practice	102
10	Plotting aids	105
	Appendix A	119
	Appendix B	127
	Index	128

Grateful acknowledgement is made to the following for per-mission to reproduce photographs: S. Smith & Sons (England) Ltd., Kelvin Hughes Division (pages 105, 106 & 114); Decca Radar Ltd. (page 112); the Marconi International Marine Co. Ltd. (pages 16 and 118); Autoplot Limited, 233 Lower Mort-lake Road, Richmond, Surrey (page 115).

Acknowledgement is also made to the Contoller of Her Majesty's Stationery Office for permission to reproduce Crown Copyright material, being the Ministry of Transport Notices, No. M.445, 'Navigation with Shipborne Radar in Reduced Visibility' and No. M.463, 'The Use of Radar'.

Introduction

The advent of radar, heralded by over-enthusiastic reports, led many to believe that dangers due to fog would be eliminated and that the mariner would be able to proceed much as he did in clear weather. Subsequent collisions between radar equipped vessels, however, have proved that this is not so and has even led some to doubt the value of radar as an aid to safe navigation.

The purpose of this booklet is an attempt to put radar into its correct perspective and to provide a simple guide to its correct use. No attempt has been made to deal with complex electronics or technical details involved in radar operation as these are adequately dealt with in other excellent text books.

Radar plotting, however, is treated on a practical basis which should prove adequate for revisional purposes and, it is hoped, encourage the reader to practise in clear weather.

Chapter six is based largely upon points which have been made by senior officers of the Merchant Navy and Fishing Fleets who have attended the Radar Simulator courses at Kingston upon Hull. The object in relating them is not to provide instructions, but to give food for thought and to indicate the advantages of reasoned action over haphazard manœuvres.

one

Why collisions occur despite radar

Marine radar has proved to be a great boon to the navigator as a means of penetrating fog, but the correct interpretation of the information it displays and the best way of using that information to avoid collision is perhaps more difficult than is generally appreciated. To the untrained observer, collision avoidance based on the impressions gained by merely watching the radar screen can often seem disarmingly simple and even after using radar for many years he may fail to realize how deceptive it can be.

Radar does not take the place of the navigator's eyes in fog; indeed it is little better than the blind man's stick probing out to feel if the way is clear and at times to help him find out where he is. Radar cannot tell black from white, steam from sail, port from starboard, stem from stern, conical from can or flashing from unlit marks and, just as the blind man's stick may miss small objects which can bring him to grief, so can radar mislead the navigator into thinking his way is clear.

Targets may not be displayed on the radar screen for several reasons:

(1) The shape, size and composition of the target may be such that the reflected energy is too weak.
(2) The surrounding sea may return echoes as strong or stronger than the target.
(3) The set may be improperly adjusted or out of order.
(4) The target may be in or beyond heavy rain.
(5) The target may be in a shadow sector.

The formidable list of reasons why a target may not be shown means that the radar observer, when confronted with a blank screen, can never be absolutely sure that there are no targets in the vicinity. Should he, therefore, have the misfortune to collide with an unseen target, it would be extremely difficult for him to justify a speed at that time in excess of that which would be considered moderate without radar. Collisions with unseen targets, however, make up only a small proportion of the total; in some cases collisions have

7

occurred between vessels which have been in radar contact for many miles.

When vessels do collide, irrespective of the warning they may have had of each other's approach, it is invariably found that at least one of them was going too fast immediately prior to the collision. This may result from one or more of the following errors:

(1) Failure to appreciate the distance required for the vessel to stop and in consequence maintaining speed for too long.
(2) Proceeding too fast for the range scale in use, so as to be unable to stop by the time a risk of collision has been determined.
(3) Failure to stop the engines when a fog signal is heard forward of the beam or when a close-quarters' situation is imminent.
(4) Failure to appreciate that a risk of collision does in fact exist.
(5) Failure to establish that action taken to avoid collision has proved to be unsatisfactory.
(6) Undue optimism that avoiding action can be taken at close range and acceptance of too close a passing distance.
(7) Assumption that, because of the locality, all other vessels will be on the same or reciprocal courses, and can therefore be passed reasonably close at speed.

The first three reasons seem inexcusable and the remedy fairly obvious. The other four stem from what might be called the optical illusion of the relative display and there are few who have not been

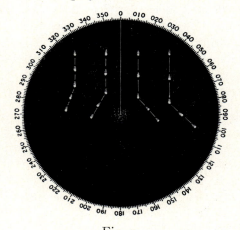

Fig. 1

misled by this at some time or other. For instance no one could determine merely by looking at fig. 1, that *all* four target vessels altered their courses 90°.

The confusion which exists between the movement of a target's echo on the radar screen and the actual course and speed of the target vessel, is perhaps the greatest factor contributing to collision. This is not necessarily because the navigator does not realize that there is a difference, but because he either fails to appreciate the importance of determining what that difference really is or he is under the impression that he can calculate it mentally. Chapter four deals fully with this and shows the necessity of keeping proper records of a target vessel's progress.

The skilled observer who by some means does manage to assess a given situation correctly and notes that there is a risk of collision, has in effect only succeeded in asking himself the question 'What is the best thing to do under these circumstances?' He has still to find the answer, although his task is much simpler than it would be for someone who had misunderstood the question.

Rules and regulations

Whenever there is movement of traffic on land, sea or in the air, it is desirable in the interests of safety to have some simple rules and regulations with which everyone must comply. When, however, owing to special circumstances, rules are necessarily imprecise and their interpretation is left in the hands of the individual, it is essential that he should exercise extreme caution.

Terms such as 'scanty information', 'close quarters situation', 'early and substantial action' may be necessary to cover all possible circumstances and conditions but this does not mean that the mariner should wait until he finds himself in some awkward predicament before attempting to consider their meaning; he should be well aware of the possible situations that may be encountered and be clear in his own mind what a reasonable interpretation might be.

Great care must be taken by the radar user who attempts to formulate his own rules. The following examples, which have been quoted by officers during discussions, are potentially dangerous if strictly adhered to even though they may have appeared to work satisfactorily.

9

(a) *I do not take avoiding action until the nearest target closes to three miles, because any action taken is more noticeable on the shorter range.* It may be true that a change of direction by a target echo shows more quickly on a short range scale, but rather like receiving a punch on the nose, there is little satisfaction in seeing it coming if there is no time to avoid it. Action taken at three miles, unless way is taken off the vessel, may leave little or no time for the navigator to change his mind should the manœuvre prove unsatisfactory.

(b) *Never alter course towards a target.* This at first sounds reasonable but, where the courses of the two vessels are crossing at a small angle, as in fig. 2, a similar alteration by both vessels away from each other could still result in collision, particularly if the alteration was small and made at close range.

Fig. 2

(c) *Avoid crossing ahead of the target vessel.* This is obviously derived from Rule 22 which is designed for situations in which only one vessel is taking avoiding action. If both vessels alter course to cross each other's stern, as in fig. 3, the result would be to speed up the collision risk rather than avoid it.

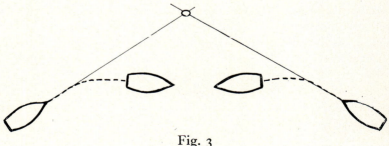

Fig. 3

Examples (b) and (c) show that the possibility of collision remains for crossing vessels if one alters to port and the other to starboard and emphasizes the necessity of making an early and bold alteration. Early, so as to allow ample time to determine that the alteration is having the desired effect; bold, so as to leave the other vessel (should it be using radar) in no doubt regarding the action taken.

two

The radar set

Although the radar display is simply a means of presenting in a visual form a succession of ranges and bearings of objects in the vessel's vicinity, radar engineers have made this information available to the navigator in a variety of useful ways. The main types of presentation are explained in the following pages. On all displays shown in this chapter, target 'A' is stationary and target 'B' is on same course and speed.

Relative unstabilized or ship's head up

This is a form of display available on all radar sets. The observer here must regard himself as being at the centre of the screen, heading in the direction of the heading marker which always points to 000°, as in figs. 4a and 4b.

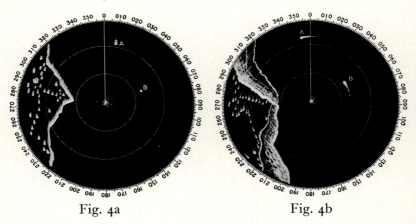

Fig. 4a   Fig. 4b

The successive positions of a target echo on this display will depend on:

(a) The target's course and speed.
(b) The observer's course and speed.
(c) Yaw or an alteration of course by the observer.

Of the three, (c) can have by far the greatest effect. For instance, a

12

vessel altering course 30° in 20 seconds would make an echo, at ten miles range, move round the screen at a relative speed of over 900 m.p.h. This would overshadow completely the effects of (a) and (b) and also result in considerable blurring of any land echoes. Even small alterations caused by yaw, make it unwise for the observer to attempt to draw any conclusions from the afterglow left by a target echo.

Bearings taken on this display are relative to the ship's head which must be noted each time a bearing is taken. This presents a serious drawback as the speed and accuracy with which bearings can be taken is an important factor in collision avoidance by radar.

Relative stabilized or north-up display

Where a gyro or magnetic repeater compass can be connected to the display it is possible to keep the picture North-upwards, as in figs. 5a and 5b. This means that all bearings will be compass

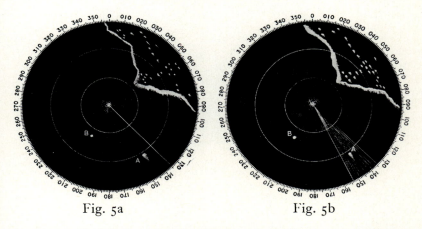

Fig. 5a   Fig. 5b

bearings and that the ship's heading marker will move round like a compass repeater indicating the course of the vessel. The observer may, therefore, take accurate bearings and at the same time keep a check on the course steered.

The movement of a target echo on this display will be a combination of:

(a) The target's course and speed.
(b) The observer's course and speed.

13

On southerly courses the heading marker will point downwards and this has caused some users to experience difficulty in co-relating what they see on the radar screen with what they would expect to see looking ahead of the vessel. To overcome this difficulty, some sets are provided with a means of rotating the picture through 180° and so enable the observer to keep the heading marker in the top half of the screen.

True motion

An alternative type of presentation available on some sets, in addition to those previously mentioned, is the true motion display,

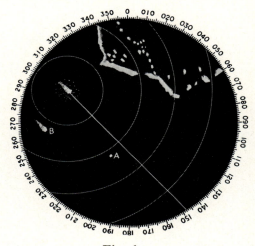

Fig. 6

as in fig. 6. Here the observer's position is made to move across the screen in accordance with his vessel's course and speed, a gyro being used for the course and the speed either fed directly from the ship's log or set manually.

Providing there is no tide and the observer's course and speed are accurately applied, the echoes of stationary targets will remain stationary on the screen, while the echoes of targets making way will indicate the target's actual course and speed, the afterglow left by an echo corresponding with the track of the target through the water.

Tide will affect the echoes of land and vessels at anchor so that they will move in the opposite direction to that in which the tide is setting the observer's vessel.

It is possible to adjust the manual controls to counteract the effect of tide and make the land remain stationary. This would mean, however, that the echoes of moving targets would indicate their movement over the ground and not the course steered. The afterglow left by an echo could now vary considerably from the vessel's track through the water, particularly when crossing a strong tide at a slow speed.

Example. If the observer was steaming north at eight knots and the tide was setting east at three knots, a target on a collision course,

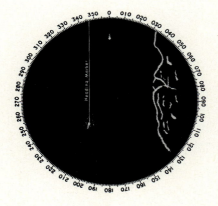

Fig. 7

bearing green 11° and steaming 202° at eight knots, would appear to be steaming south as in fig. 7. The targets aspect would, therefore, appear to be green 11° instead of red 11°; potentially, one of the worst mistakes that can be made in the use of radar.

Since the observer does not remain at the centre of the screen, the mechanical bearing-cursor cannot be used so an electronic bearing-marker similar to a rotatable heading marker is provided, the bearing being read on a separate scale.

When true motion is in use, it is not advisable to allow the observer's position to approach too close to the edge of the screen as the view ahead would be seriously restricted. The picture should be reset when the observer's vessel is about half-way across the screen,

Ship's head-up stabilized

A facility which has recently been introduced in a new type of British Radar Equipment in addition to the more common unstabilized and North up stabilized displays, is the stabilization of the display ship's head-up (fig. 8), either relative or true motion, the object being to have all the advantages of a stabilized display

Marconi's 'Argus 16' Stabilized Screen Display

but keeping the heading marker pointing upwards to avoid the observer having to perform mental gymnastics.

Picture quality

Whatever type of display the navigator may use, it is important that he should be able to determine that the set is functioning correctly, particularly when no echoes are showing on the screen.

One means of checking the performance is by using what is called a *Performance monitor* or *Echo box*. This is a metal box which will resonate (ring) at the radar frequency when struck by energy from the transmitter. While it is resonating it is re-transmitting energy to the radar receiver which displays it in the form of a plume (fig. 9) or on some displays as a sun effect. The length of the plume or radius of the sun should remain constant when the set is working satisfactorily, a decrease in length indicating a reduction in performance.

The length of the echo produced by the performance monitor should be measured when the set is working satisfactorily and its length together with the control settings at the time of measure-

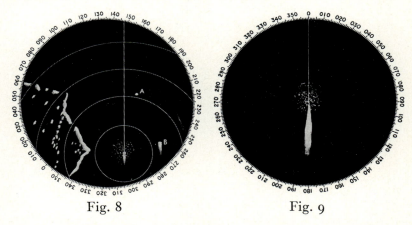

Fig. 8   Fig. 9

ment should be entered in the radar log or scrap log for future reference.

Frequent use of the performance monitor is the only sure method of discovering any slight deterioration in performance or maladjustment of the controls. Measurement of the extent of sea-clutter, even when it is present, is a poor second best as an indication of picture quality.

If log entries are made regarding the plume length each time the set is used, these records will provide useful proof that the set was in correct adjustment should the user be unfortunate enough to collide with an unseen target.

Although it is possible for anyone switching on a radar set and turning all the knobs in sight to obtain a picture of sorts, this is obviously not good practice. In order to obtain the best picture the set is capable of providing, the user should carefully follow the maker's instructions, although the following procedure should prove satisfactory for most sets:

(1) See that the aerial is clear of all obstructions.

(2) Turn the Brilliance, Gain (Sensitivity) and Anti-clutter off.

(3) Set the pulse length switch to 'Long', if possible.

(4) Set the rain switch to 'Normal', if so marked.

(5) Set the 'Stand-by/Transmit' Switch to 'Transmit'.

(6) Select a long or medium range on the relative unstabilized display.

(7) Switch on and wait about three minutes for the transmission to start.

(8) Turn up the brilliance control until the rotating trace is just visible.

(9) Switch on the range rings and adjust the focus control by slowly rocking it across the point of minimum thickness of the rings, gradually reducing the movement until the thinnest lines possible are obtained.

(10) Turn off the range rings and adjust the brilliance so that *the trace just disappears.*

(11) Check that the centre spot is directly under the centre of the bearing cursor. This can be done by switching on the electronic bearing marker and setting it at 90° from the heading marker, then seeing that the mechanical bearing cursor will lie directly over them both in turn. Where the electronic marker is not fitted, the heading marker may be used on both 000° and 090°.

(12) Reset the heading marker to 000°.

(13) Turn up the Sensitivity (Gain) so that a faint speckled background extends to the edge of the display.

(14) Turn to the 3-mile range and switch on the performance monitor. Adjust the manual tuning control (where fitted) for maximum length of plume. If a performance monitor is not fitted, then tune for maximum sea-clutter. Some sets have a neon light or meter to facilitate tuning.

(15) Switch off the performance monitor and adjust the anti-clutter control so that sea echoes are just painting on the screen (using the short pulse length).

Note. The anti-clutter control is very critical. If too much is used, targets as well as sea echoes may be removed and, if too little is used, then target echoes will not be seen against the bright background of the sea-clutter. It must be emphasized, however, that the anti-clutter control does not possess any magical powers and any targets which return weaker echoes than the sea surrounding them cannot be made to appear.

(16) Check for correct number of range rings and that they are evenly spaced.

The fact that the radar has been correctly adjusted and is working satisfactorily does not in itself mean that the operator can make full use of the set as a safe aid to navigation. He must also know to what extent he can rely on it to display possible hazards. This will depend on:

(1) State of the weather and sea. (2) Height of the scanner.
(3) Transmitting power. (4) Reflecting properties of the targets.

Undoubtedly the best way to achieve and maintain confidence in the radar is to have frequent practice in clear weather. In this way comprehensive records can be built up of the ranges at which various types of targets are first displayed, together with the quality of the radar picture at the time. It is recommended that a special radar log should be kept for this purpose and entries in clear weather might include:

(1) Date and place. (2) Description of the target.
(3) Detection range. (4) State of weather and sea.
(5) Length of performance monitor plume.
(6) Unusual echoes or defects in performance.
(7) Repairs and replacements.

When the radar is in use owing to restricted visibility or the possibility of it occurring, the following entries in the scrap or radar log would assist in avoiding any misunderstandings when the radar watch is changed:

(1) Date and time. (2) Ranges and bearings of targets ahead.
(3) Estimated nearest approach of targets already plotted.

(4) Speed by log.

(5) Length of the Performance Monitor plume.

(6) State of weather and sea.   (7) Visibility.

(8) Range scale in use when watch changed.

three

Miscellaneous effects and errors

Shadow sectors

Just as the observer's view of the horizon in clear weather is interrupted by the ship's superstructure and other objects in the vicinity, so is that of radar, and targets which are wholly or partially screened from the radar scanner are said to be in shadow sectors.

The extent of shadow sectors may be determined by turning the vessel through 360° when a buoy shows on the screen, and noting the relative bearings when it begins to disappear and when it returns to maximum brilliance. The absence of sea-clutter may also provide

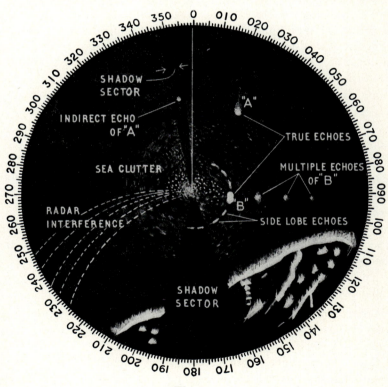

Fig. 10

an indication of the extent of shadow sectors. These sectors should be entered in the radar log, but it should be borne in mind that they could be affected by a change of trim or stowing the derricks vertically.

Shadow sectors ahead of the vessel are naturally dangerous and periodic alterations of course should be made, in order to scan them.

Indirect echoes

Radar energy deflected by objects causing shadow sectors may strike other targets and return by the same path. If the returning echo is strong enough it will be displayed as a false echo in the shadow sector as in fig. 10.

Multiple echoes

When a target is close to the observer's vessel, the transmitted radar energy may bounce to and from between his vessel and the target until it is dissipated. If, on each return journey, some energy reaches the radar receiver, the effect on the radar screen will be a succession of echoes on the same bearing at intervals equal to the target's range. The nearest echo will be at the correct range of the target.

Side-lobe echoes

Although most of the transmitted radar energy is confined to a narrow beam or lobe of about $2°$ or $3°$ width, small amounts (called side-lobes) spread out on either side of the main beam. Large targets close to the observer's vessel may return echoes from these side-lobes which will show as an interrupted arc of up to $90°$ on either side of the true echo. Side-lobe echoes may extend across the shadow sectors.

Radar interference

When radar waves transmitted by other vessels are at the same frequency as the observer's radar, they may be displayed on his screen as a pattern of dotted lines.

Radar interference is not necessarily an indication that a target vessel showing on the screen is using radar; it may be caused by some vessel beyond the observer's range. Neither does the absence of radar interference mean that the targets which are showing on the screen are not using radar. They may be using a different frequency.

Non-standard propagation

The normal radar range can be extended or decreased by atmospheric effects. In general, one may expect a decrease in range when the air is colder than the sea, as in Polar regions, and an increase in range when the air is warmer than the sea, as in the Mediterranean type of climate. Although other factors such as the relative humidity play their part, the navigator would unlikely be able to determine or anticipate their effects. (See Refraction.)

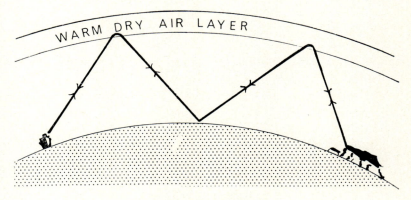

Fig. 11. Ducting

Freak reception

Freak reception is sometimes obtained as a result of the radar waves bouncing round the earth's surface between a warm air layer and the sea. This is known as *ducting*, (fig. 11) and can result in echoes being received from well outside the maximum range of the set.

Second-, third- and fourth-trace returns

The range at which a target echo is displayed on the radar screen will depend on the time interval which elapses between a pulse of radio energy being transmitted and the echo's return. Under freak conditions, an echo may return from a distant target after a second pulse has been transmitted. The set will then display this echo as if its origin was the second pulse. In this way, it is possible for an echo to be displayed at, say, 10 miles when the target's range is really 130 miles. Under exceptional conditions it is possible for echoes to return after three or even four pulses have been transmitted, these being called third- or fourth-trace returns. Land echoes seen at close range (when the land is actually some considerable distance away), could be most disconcerting for the navigator who does not realize that this type of echo can occur.

As collision avoidance and position-fixing by radar must depend on the accuracy with which ranges and bearings are measured, the navigator must be fully conversant with the possible errors which may affect them.

Bearing errors

These are described as they would apply to the 'Relative Unstabilized Display', although most are applicable to the other types of display.

(1) The heading marker must be accurately set to 000° or all bearings will be in error by the amount it is displaced.

(2) If, when the heading marker flashes on at 000°, the scanner is not pointing right ahead of the vessel, targets will be displayed incorrectly by the amount it is out of alignment. This error can be checked by comparing radar and visual bearings of a distant target taken simultaneously. (See M.O.T. Notice M.535.)

(3) The centre of the rotating trace should be directly under the centre of the mechanical bearing-cursor. Any displacement would produce an error, as indicated in fig. 12.

This error increases as the target approaches the centre of the screen and so it is advisable to use the shortest range scale possible when taking bearings.

(4) As the mechanical bearing-cursor is some distance from the face of the radar screen, there is a possibility of an error from parallax unless the observer keeps his head well over the display.

(5) True bearings are obtained from the 'Ship's Head Up' display by applying the relative bearing to the ship's true course. It follows that if an incorrect compass error is applied to obtain the true course this will also affect all the bearings.

24

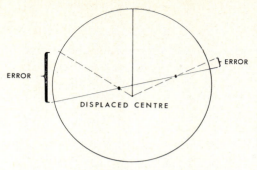

Fig. 12

(6) Errors may be caused by the vessel being unknowingly off course owing to yaw, etc.
(7) Bad focusing can give extra width to targets.
(8) Radar energy goes out in a beam, and echoes returning from the edge of the beam are displayed as if they were returning from the beam's centre line. This has the effect of extending the edge of a coast line for half the beam's width, as in fig. 13. The width of the beam can be reduced by turning down the gain, but this should not be attempted unless the echoes are very strong.

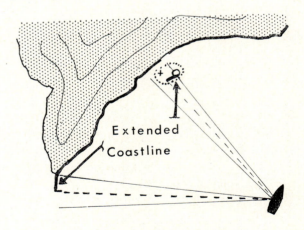

BEAM WIDTH DISTORTION

Fig. 13

Range errors

(1) An uncalibrated variable range marker.
(2) Wrong interpolation when using range rings.
(3) Error inherent in the equipment as indicated by the makers.
(4) Poor focusing, giving enlarged targets and thick range ring.

Factors affecting the production of radar echoes

R ange	**E** nvironment
A rea	**C** omposition
D uration of pulse	**H** eight
A spect	**O** bstruction
R efraction	**E** fficiency
	S urface

Range

The strength of an echo returning from a target will increase considerably as the range is reduced. If, for instance, the range is halved the echo's strength will increase to about sixteen times its original value.

Area

The amount of energy returned from a target will increase in proportion to the reflecting area, the upper limit being the cross-sectional area of the radar beam.

Duration of pulse

The longer the radar pulse, the greater the amount of radar energy transmitted and an echo is therefore more likely to be returned from a long range. Short pulses give better range discrimination, i.e. two targets close together on the same bearing are more likely to be displayed separately.

Aspect

The angle between the reflecting surface of a target and the line of transmission will determine the direction in which the radar energy is reflected.

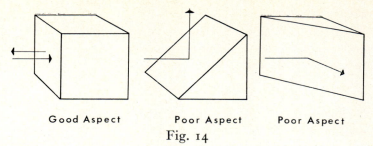

Good Aspect   Poor Aspect   Poor Aspect
Fig. 14

Refraction

The range of the radar horizon will depend upon the amount that the radar beam is bent in passing through the atmosphere. This bending is caused by the change in the temperature and relative humidity of the atmosphere as the height above the earth's surface increases. A greater than normal decrease in temperature and an increase in the relative humidity with height will give reduced bending (Sub-refraction), whilst a less than normal decrease in the temperature or a reduction of relative humidity with height will give greater or Super-refraction.

Sub-refraction causes a reduction in the normal range whilst super-refraction causes an increase.

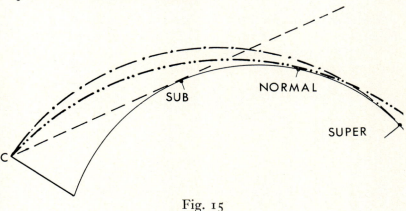

Fig. 15

Environment

Echoes returning from the area surrounding a target may be so strong as to make it difficult or impossible to pick out the target. These may be echoes from land or sea-clutter.

Composition

The ability of the target to reflect or re-transmit radar energy will depend on its electrical qualities, good conductors of electricity being normally the best reflectors.

Height

The maximum range at which a target may be detected will depend on both the height of the scanner and the height of the target, the range for normal atmospheric conditions being approximately:
$$1\cdot22\sqrt{\text{height of scanner}} + 1\cdot22\sqrt{\text{height of target}}.$$

Obstructions

The ship's superstructure, land, heavy rain or snow may screen a target from the scanner. (See shadow sectors.)

Efficiency

Good response from the radar will depend on the condition of the set and the efficiency of the operator.

Surface

When the aspect of a target is good, a smooth surface will reflect the most energy but, when the aspect is poor, a rough surface is more likely to return some energy to the scanner, e.g. sea-clutter increases when the sea becomes rough.

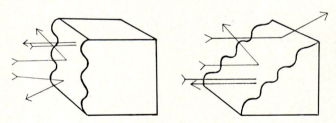

Fig. 16

Weather Effects
Rain

Rain shows on the screen as a bright area, the intensity of which depends upon the amount of precipitation. Very heavy rain may completely obliterate other targets and create a shadow sector beyond the rain.

It is sometimes possible to detect strong targets in rain by *a temporary* reduction of the gain. This reduces the effect of rain echoes whilst the stronger targets remain bright and show against the darker background.

The rain switch or differentiator may also help to detect targets in rain, as it has the effect of allowing only the leading edge of constant strength echoes to show on the radar screen. This means that although the nearest edge of the rain will show, the rest of the signal will fade, but if there is a *stronger* signal returned from within the rain it may now show better against a darker background.

Snow

Snow being less dense than rain does not produce such strong echoes, but when composed of large wet flakes it can seriously impair the radar picture. An accumulation of ice and snow on the radar scanner will also have a serious effect and it should be kept clear.

10 cm Radar

Radar sets using 10 cm rather than the normal 3 cm wavelength are much less affected by rain and snow.

Icebergs

The reflecting properties of icebergs are normally less than half that of a ship of similar size; when its aspect is poor an iceberg can be difficult to detect. Sea-water has a much greater response than ice and in consequence in a choppy sea it is quite possible that growlers large enough to do damage may not be seen on a correctly adjusted radar display. On the other hand in calm water small pieces of ice can have an alarming appearance on the radar screen.

Sea Clutter

Because of the shape of the waves, sea clutter will usually be much stronger on the windward side of the vessel. It follows that poor targets will show better at close range if to leeward of the vessel.

Sea clutter can extend to about three miles and the higher the scanner the greater its effect.

Corner Reflector

Considerable improvement in the radar reflecting properties of small targets can often be obtained by using a corner reflector as in fig. 16a.

This is simply three metal plates mutually at right angles to each other. Radar energy striking one of the corner sections will be reflected back in the reverse direction irrespective of the angle of impact.

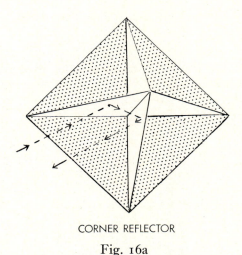

CORNER REFLECTOR

Fig. 16a

four

Radar information and its application

Examination of radar collision cases invariably shows that the failure to record the progress of target echoes has resulted in the situation being misunderstood. Even so, some mariners remain unconvinced of the necessity of keeping records or feel that they have insufficient time when there are several echoes showing on the screen. This is quite illogical. A blind man does not drag his guide dog across a busy road; he waits for a clear indication that it is safe to cross. It may be true that if cars were as far apart as ships and travelled at the same speed, he could frequently make a successful crossing, but his first mistake may prove fatal.

The mariner who is unable to keep some graphical record of the movement of target echoes is probably untrained, out of practice or going too fast. Whatever the reason, it can be no excuse for taking action on guess-work or proceeding faster than he would without radar. On the contrary, in such circumstances, he should probably be going much slower.

The movement of target echoes

The first thing the relative display user must realize is that the track of a target's echo across the radar display does not necessarily indicate the direction in which the target is steaming. In fact, the only time the echo's movement will coincide with the target's actual course and speed is when the observer's vessel is stopped. See figs. 17a and 17b.

When the observer's vessel is making way, a stationary target will appear to move in a direction parallel and opposite to his course. The distance moved by the target is the same as that steamed by the observer's vessel as in figs. 18a and 18b.

When both the observer and the target vessel are making way, the track of the echo across the screen will be a combination of both vessels movements, see figs. 19a and 19b.

Apparent motion line and nearest approach

The line drawn through successive positions of a target echo as

31

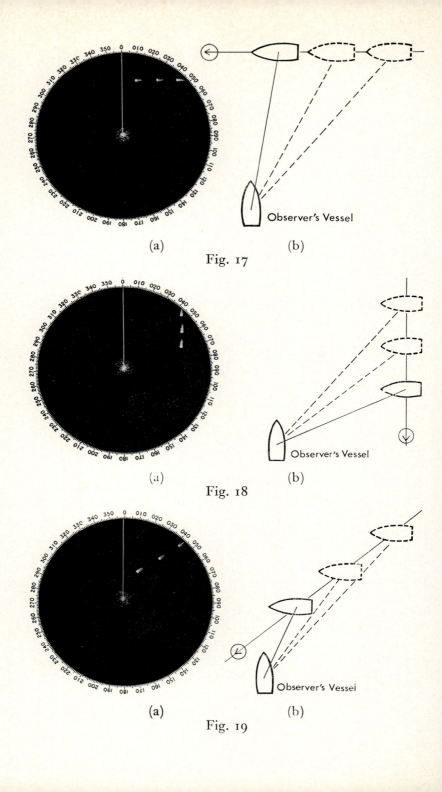

(a)   (b)

Fig. 17

(a)   (b)

Fig. 18

(a)   (b)

Fig. 19

seen on the relative display is called the Apparent Motion Line and is usually denoted thus ⊖. This line is of prime importance in determining the nearest approach and the time of nearest approach of the target vessel.

In fig. 20, a line drawn through the three positions of the target taken at 5-minute intervals: O, A, and A^1, gives the apparent motion line OX. The point of nearest approach is found by drawing a line from the observer's position to cut the apparent motion line at right angles. In the figure the nearest approach will be 1 mile, assuming that both vessels keep their course and speed.

As the target echo is moving across the screen 1 mile every 5 minutes and the distance from A^1 to X is 4 miles, it follows that if

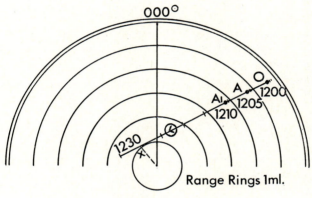

Fig. 20

the target was at A^1 at 1210 it will arrive at X, 20 minutes later, at 1230.

When the apparent motion line passes through the centre of the screen, it means that the target is on a steady bearing and the nearest approach will be collision as in fig. 21.

Error in estimation of nearest approach caused by inaccurate bearings

To enable early avoiding action to be taken, any risk of collision must be determined at a reasonably long range. This will require accurate bearings as a small error at long range can mean a large error in the nearest approach.

RW–C

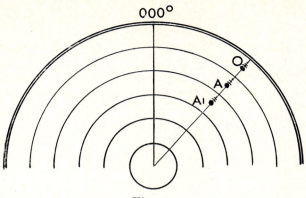

Fig. 21

For example: fig. 22 shows a target which was first detected at 9 miles and is approaching on a steady bearing. If, however, an error of 3° was made in the bearing when its range was 6 miles, it could seem to the observer that the target's nearest approach would be 1 mile on either the port or starboard side. This error could have considerable influence on the action the observer may decide to take.

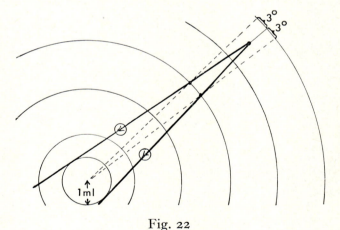

Fig. 22

Effects of the target altering course

A change in the direction in which a target echo is moving may mislead the radar viewer into believing that the target has altered course a similar amount, but this can be far from the truth. The change in direction may be due to either the observer's vessel or the target altering course or speed but, even when the target has altered course, the amount may vary considerably from the change in apparent motion.

Consider a target vessel meeting the observer end on, the target steaming at 5 knots and the observer at 15 knots.

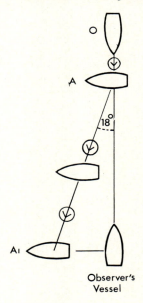

Fig. 23

If the target then alters course 90° the change in the apparent motion will be as in fig. 23.

At O the target will have a relative movement of 20 knots towards the observer (the sum of both speeds).

At A the target alters course 90° to starboard, say at six miles ahead of the observer. By the time the observer has steamed the 6 miles to bring the target abeam, the target will only have steamed 2 miles to A^1.

It can be seen from the figure that the resulting change in the apparent motion line is only about 18°. It is of interest to note that, if both vessels had been proceeding at the same speed, the change in apparent motion would be half of the alteration of course by the target vessel.

Fig. 24 shows the tracks of four vessels originally on opposite courses to the observer. All the targets were steaming at 8 knots and the observer at 16 knots.

The four targets then altered course 15°, 30°, 60° and 90° from left to right respectively. The change in apparent motion caused by the alterations of 15° and 30° are so small that they may not be noticed for some time, particularly when affected by yaw.

There are two morals to learn from this:

(1) When the observer's speed is excessive it is most difficult to detect avoiding action taken by the target vessel.

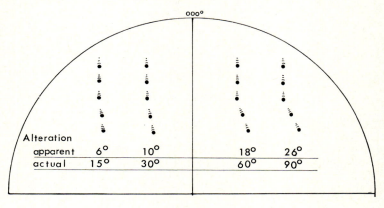

Fig. 24

(2) A bold alteration is absolutely essential if it is to be noticed by the other vessel, particularly if she is proceeding at a faster speed. 'Bold' meaning not less than 60°.

The foregoing is also the explanation of the target echoes movements in fig. 1, chapter one, the targets' speeds being, from left to right, 6, 10, 14 and 18 knots respectively, whilst the observer was steaming at 18 knots.

The steady bearing

In clear weather, when another vessel is closing, it is common practice for the navigating officer to stand in one particular position on the bridge and note the bearing of the other vessel relative to some part of his ship's superstructure; his intention being to stand in the same position later and see if the bearing has remained steady. *He would only do this if the other vessel appeared to maintain the same course.* It is this latter point which is sometimes forgotten by the radar observer who puts the bearing cursor on a target echo and looks again later to see if the target is still on the cursor line.

Fig. 25 shows a vessel which, although it could have passed ahead, has decided to alter course to starboard. If the radar observer only looked at his screen when the target was at O and A^1, he would think that the target was on a steady bearing.

Fig. 26 shows a similar case in which the target vessel has stopped.

Both these examples clearly indicate that it is essential to take

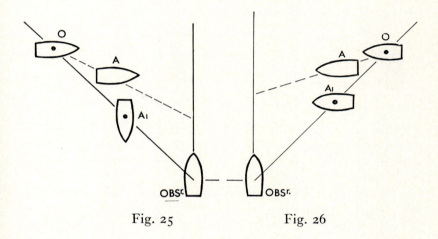

Fig. 25   Fig. 26

more than two bearings before coming to any conclusion as to the risk of collision and before taking any avoiding action.

The changing bearing

In this example let us consider the case of a target vessel not equipped with radar or whose radar has broken down. If the target vessel is stopped when it is first seen (possibly because of hearing the whistle of some vessel not on the observer's screen) and it then gets under way as the observer's vessel approaches, the result could be as indicated in fig. 27.

Here the bearing is constantly changing and if the observer is only watching the bearing and not plotting the target's position the risk

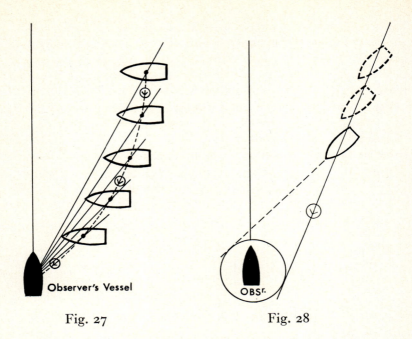

Fig. 27   Fig. 28

of collision may not be appreciated until too late. Only by a continuous graphical record can the curve of apparent motion be appreciated.

This situation could be particularly dangerous if the observer, believing he was passing clear, neglected to sound his fog signal. The target vessel would not then be aware that there was anything ahead.

A dangerous assumption

When two vessels are meeting, as in fig. 28, the apparent motion of the target tends to give the impression that the vessels are on opposite courses. This impression, when coupled with the fact that the vessels are on a route not crossed by other traffic lanes, is often accepted too readily and the observer makes no effort to find out what the target's course actually is.

Here we have the opening gambit for the typical collision case, which, although it has been described in several publications during past years, and regularly demonstrated on radar simulator courses, continues to be extremely popular.

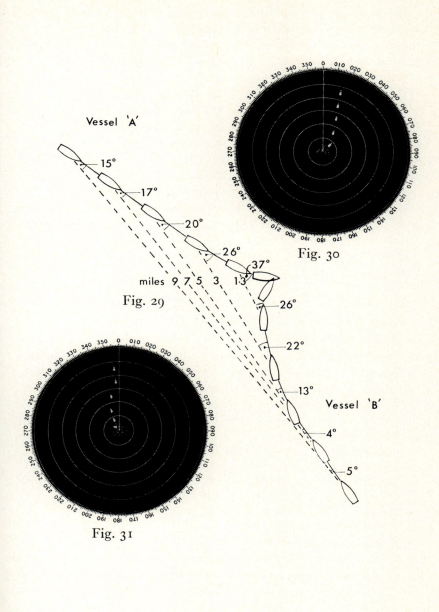

The typical collision case

Fig. 29 shows two vessels approaching in a similar manner to that previously described, each steaming at 10 knots. The insets (figs. 30 and 31) show the successive positions of each other's echoes as seen on their respective radar screens.

A's interpretation of the situation (fig. 30)

Target B was first seen 15° on the starboard bow, range 9 miles and the bearing opened to 20° on the bow by the time B was 5 miles away. From this A inferred that the vessels would pass each other, starboard to starboard. When B's range was 3 miles, A altered course 5° to port to increase the passing distance but by the time B had approached to 2 miles, this appeared to have had little or no effect and A altered another 5° to port. When the vessels were 1 mile apart, it was evident that B was going to pass much closer than was expected and a further alteration of 20° to port was made, but B continued to close and appeared on the starboard beam when the collision occurred.

B's interpretation of the situation (fig. 31)

Vessel B first saw target A 5° on the port bow and came to the conclusion, after watching the echo approach to 5 miles, that it was an almost end-on case. He therefore made an alteration of 10° to starboard. As the target echo still appeared to close, a further two alterations of 10° to starboard were made. Each time, although the relative bearing opened, target A continued to close. At 0·5 miles B, realizing that something was wrong, put his helm hard to starboard, unfortunately too late.

The faults which are usually found in this type of collision case are:

(1) Proceeding at an immoderate speed.
(2) Making false assumptions on scanty information.
(3) Failing to take early and *substantial* action to avoid a close quarters' situation.
(4) Not stopping when a fog signal was heard forward of the beam, or when a close-quarters' situation was imminent.
(5) Making a succession of small alterations of course instead of a single bold alteration.

five

Radar plotting

The target's course

It should be evident from chapter four that, in order to estimate the nearest approach of another vessel, it is essential to keep a continuous record of its track across the radar screen. Once this is done the extra work involved in finding the target's course is so little and yet so important, that it would be extremely foolish to omit it.

It will probably assist the newcomer to the relative plot if he imagines that, at the time a target is first observed, it is passing a buoy. Now, although it is impossible to forecast the future position of a moving target, the navigator does know that the buoy (neglecting tide) will move parallel and opposite to his own course. If he then plots the estimated position of the imaginary buoy for the time he will next observe the target, any difference between their positions must be due to the target's own movement.

Fig. 32 shows three successive positions of a target and the corresponding positions of a buoy which the target passed at O. It can

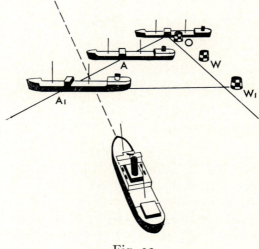

Fig. 32

be seen that, when the observer has steamed so as to bring the buoy to W, the target has steamed from W to A. Similarly in the time it has taken for the buoy to move from O to W¹, the target has steamed from W¹ to A¹.

Aspect

Rather than calculate the target's true course, some masters prefer to use the angle between the direction a target is heading and a line joining both vessels. This is the target's 'Aspect'. Fig. 33 shows a target having an aspect of Red 30°. It is in fact the target's relative

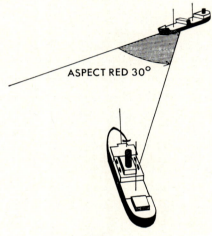

Fig. 33

bearing of the observer, measured from the target's course line up to 180° on either side. It is called red if the target is presenting its port side and green if the starboard side.

THE SIMPLE PLOT

In making a simple plot, it is essential to take into consideration the distance covered by the observer's vessel during the plotting interval. Some navigators prefer to use a fixed time interval of 6 minutes as the distance run can then be found by simply dividing the ship's speed by 10. For fast vessels, it would be necessary to take an intermediate observation after 3 minutes. Other navigators prefer to use a fixed run of 2 miles on the 12-mile range with an intermediate observation after 1 mile. On the 6-mile range, a one-mile run would be used with an intermediate observation after 0·5 miles. The latter method automatically gives quicker results for faster vessels and at the same time gives an adequate run for reasonable accuracy.

The following example shows how the target's course would be found in practice. It is done in three stages as it would be at sea although only one plotting sheet would be used.

The plotting diagrams have been constructed as the targets and heading marker would appear on a 'North-up' display, as this method will give continuity of the target echoes' movement when the observer alters course. It will be appreciated that when a 'Ship's head up' display is in use, the target's bearing must be laid off relative to the ship's course or converted to a compass bearing.

The simple plot

First observation. Time, 1200 hours. Observer's course and speed, 330°, 10 knots.

Procedure

1. Measure the range and bearing of the target and note the time or log reading.

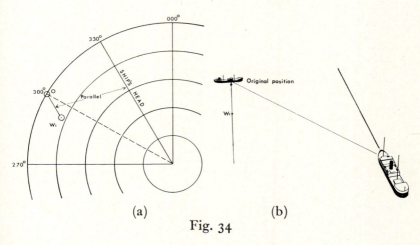

(a)   (b)

Fig. 34

2. Lay off the true bearing and range of the target on the plotting paper marking it O for the original position.
3. Draw a line from O parallel and opposite to the heading marker marking a position 2 miles from O and name it W^1. If we imagined the target passed a buoy at O the buoy would be at W^1 after the observer had steamed for 2 miles.

Report—Target bearing 30° on the port bow, range 10 miles.

The time required to take the radar range and bearing should not exceed that required to take a visual bearing.

Second observation. Time 1206 hours. Observer's course and speed, 330°, 10 knots.

After steaming for 1 mile (or other selected unit), put the bearing and range of the target on the plotting diagram again. Name it A. If there is more than one target they should be called A, B, C, etc. *Note.* No assumptions should be made at this stage regarding the course and speed of the target but, if the target is closing fast, it may be advisable to shorten the interval before taking the next range and bearing.

Report—Target still bearing 30° on the port bow. Range has closed 1·5 miles in 6 minutes and is now 8·5 miles.

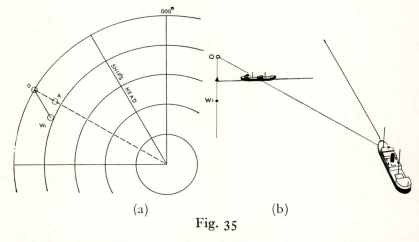

(a)   (b)
Fig. 35

Third observation. Time 1212 hours. Observer's course and speed, 330°, 10 knots.

Procedure

1. After steaming another mile, plot the target's position again and mark it A^1.
2. Join W^1 to A^1 and this is the target's course and distance steamed in the interval providing that O, A and A^1 are in line and equal distances apart.
3. The target's course may be found by drawing a line parallel to $W^1 A^1$ through the centre of the plotting diagram to the degree scale.
4. To find the target's speed in knots measure W^1 to A^1 and multiply it by the ratio of 60 minutes to the plotting interval in minutes. In this case multiply by $\frac{60}{12}$ as the plotting interval was for 12 minutes and we require the distance steamed in 1 hour.

$$\text{Since } W^1 \text{ to } A^1 = 1 \cdot 6 \text{ miles}$$

$$\text{the target's speed} = 1 \cdot 6 \times \frac{60}{12}$$

$$= 8 \text{ knots.}$$

Report—Target closing on a steady bearing 30° on the port bow.
Target's speed—8 knots approximately.
Aspect—Green 40° or (Course 080° T.).
Range—7 miles.
Risk of collision—in 28 minutes at 1240 hours.

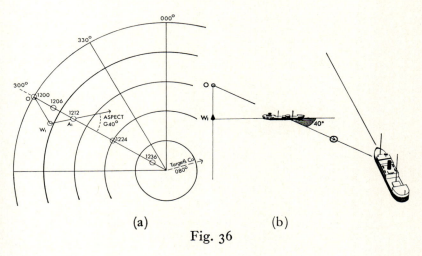

(a)   (b)

Fig. 36

Avoiding action

The observer is now in a position to decide on the most appropriate avoiding action, bearing in mind that the target may also take action at the same time.

The ideal manœuvre is one that would make it impossible for the target to re-introduce a collision risk, *whatever it may do*. Unfortunately, this is not always practicable and the choice must rest on one that offers the best chance of success.

In addition to considering what the target vessel may do, the observer must be sure that his own action will avoid a close-quarters' situation if the target keeps its course and speed. Manœuvres made without a clear idea of how these will affect the target's apparent motion, mean that the observer is living on his nerves, waiting to see what will happen. Frequently in such cases, the change in the target's apparent motion is not what was expected and the impression is gained (quite wrongly) that the target has also taken action. This can lead to quite unnecessary complications.

Undoubtedly the easiest solution to visualize is the effect of stopping the vessel as the movement of the echoes will then be the same as the target's course and speed. Once the simple plot is completed and the target's course is known, the observer should be able to tell almost at a glance, if stopping would avoid a close-quarters'

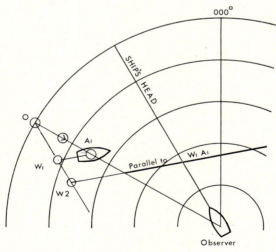

Fig. 37

situation. The only complicating factor is the distance required for the vessel to stop, a factor which is frequently under-estimated. (See Coastal navigation.)

Advisability of stopping

Fig. 37 shows how the observer may consider the advisability of stopping. All lines shown, other than for the simple plot, are not required in practice by the experienced plotter.

Procedure

1. Estimate the distance that the vessel will travel before coming to rest since the last observation. This would be the distance an imaginary buoy at W^1 would move to W^2.
2. A line from W^2 parallel to the target's course, W^1A^1, will be the target's track (unless it alters course). If this line passes too close to the observer's position at the centre of the plot, then some alternative action should be considered.

If it should be decided to stop, then the plot should be continued as in fig. 38 (for the fourth observation).

Procedure when stopping

Fourth observation

(1) Having decided to stop, the target's position should be

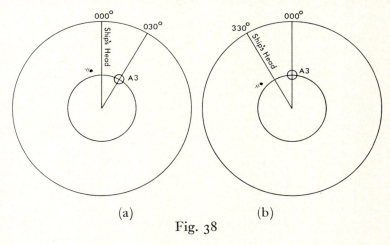

(a)   (b)
Fig. 38

plotted when the observer's vessel is no longer making way (A^2). From this position the target's echo on the radar screen will move in the same direction and at the same speed as the target vessel's actual course and speed.

(2) When the target maintains its original course and speed, the echo will move from A^2 parallel to W^1A^1 shown as A^2A^3 in fig. 38c.

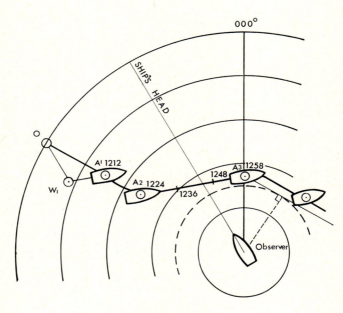

Fig. 38c

Resuming speed

(1) Before resuming speed the observer must decide upon an appropriate minimum range for the target to pass, and be reasonably sure that the target will remain outside this range (3 miles is used in this example).

(2) If the target continues to keep the same course and speed, when the observer resumes his own speed, the echo will return to its original direction and rate of apparent motion, OA^1. The observer may therefore proceed when the target's echo reaches A^3 on a line drawn parallel to OA^1 which just touches the 3-mile range ring selected for the target's nearest approach.

(3) Since W^1A^1 is the distance travelled by the target in 12 minutes, the observer can check that the target keeps its speed by marking on A^2A^3 a succession of distances equal to W^1A^1, thus obtaining the positions the echo should be in at 12-minute intervals and the time the target will arrive at A^3.

(4) A simple method of determining when to resume speed is to set the bearing marker (electronic if fitted) to the bearing of A^3 and the variable range ring to its range (see Radar diagrams on page 47). If the target moves to this point it must be keeping its course and if it arrives at the correct time it is also maintaining its speed. The observer should then be able to proceed, the time taken to regain the original speed being an additional safety factor.

Altering course

The change in a target's apparent motion caused by the observer altering course is difficult, if not impossible, to visualize from casual observations of the radar screen, but fortunately it can easily be found from the simple plot.

In the following explanation of the principle on which this is based, use has again been made of an imaginary buoy which the target is assumed to have passed when it was first seen at 1200 hours.

In fig. 39, A^1 and W^1 represent the positions of the target vessel and the imaginary buoy at 1212 hours. If the observer's vessel has been steaming 330° T. since 1200 hours at 10 knots, the target and the buoy must have originally been together at O (2 miles, 330° T. from W^1). The target's apparent movement being from O to A^1 (3 miles in 12 minutes) if the observer remains on the same course a further 12 minutes, the target will move another 3 miles to X.

Suppose, however, that the observer has been steaming 030° T. at 10 knots since 1200 hours. In this case the original position of the imaginary buoy would have been at O^1, 2 miles, 030° from W^1, and the apparent motion of the target from O^1 to A^1 (1·5 miles in 12 minutes). If the observer continues to steam 030° T. for another 12 minutes, then the target would move to Y, 1·5 miles from A^1.

It should be clear from the above that, if the observer altered his course from 330° T. to 030° T. when the target was at A^1, the target's apparent motion would change from OA^1 to A^1Y. The new direction and rate of a target's apparent motion may therefore be found by plotting the position of O^1 which is the position the target would have been in at the beginning of the plotting interval, assuming the observer had been on the new course since that time. O^1 to A^1 is the desired new rate and direction of apparent motion.

50

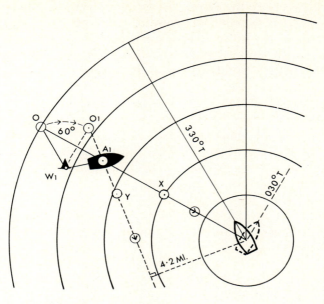

Fig. 39

Method of investigating the merits of altering course

(1) When the simple plot has been completed, rotate O about W^1 by the amount it is proposed to alter course to O^1. ($W^1 O^1$ will now be parallel to the new course line.)

(2) From O^1 draw a line through A^1. This is the target's new apparent motion line from A^1, assuming the observer altered course instantaneously when the target was at A^1.

(3) The distance from O^1 to A^1 is the amount the target will move across the screen in a time interval equal to that which was used in the simple plot.

(4) In deciding if the new apparent motion line would be satisfactory, an allowance must be made for the time which would

elapse before the vessel was steady on the new course. This allowance should be known from previous experience (*practice in clear weather is desirable*).

Should it be decided to alter course 60° to starboard, the experienced plotter could anticipate the future movements of the target's echo as follows:

Fourth observation at 1224 hrs

1. If not already done, mark the position of O^1, 2 miles, 030° T. from W^1. Join O^1 to A^1. This will be the new direction and distance the echo will move in 12 minutes (fig. 40).
2. When steady on the new course 030° T. re-plot the target's position A^2.
3. From A^2 draw A^2Y parallel to O^1A^1. The echo should now move along this line, unless the target vessel alters its course or speed.

Resuming course

When the observer resumes his course, the target's apparent motion should revert to its original direction and speed, i.e. OA^1.

4. Having selected the range ring for the desired nearest approach of the target (in this case 3 miles), draw a line parallel to OA^1 so that it just touches this range ring. This line will cut A^2Y at A^3.
5. The time the target should arrive at A^3 (1256) is found by dividing A^2A^3 into 12-minute units, using the distance O^1A^1.
6. A useful indication of when to resume course is to set the variable range marker and bearing marker to the position of A^3. The original course being resumed if the target arrives at A^3 at the estimated time.

The delay in regaining the original course will slightly increase the passing range.

Reducing speed

A small reduction in speed, even though it may be all that is necessary in order to avoid a close-quarters' situation, is probably the least satisfactory type of manœuvre. In addition to the difficulties involved in assessing a situation when the observer's speed is

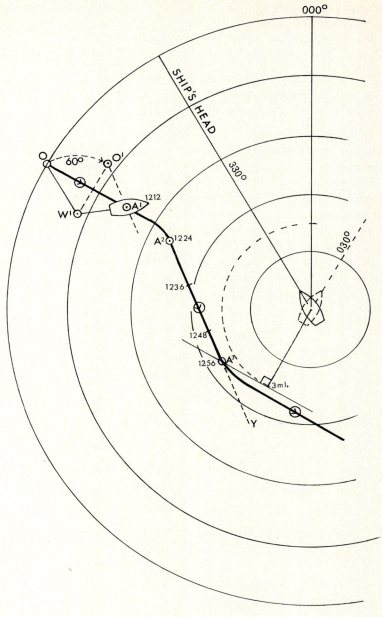

Fig. 40

changing, it is obviously impossible to estimate the effect of a reduction if the ultimate speed is in doubt.

When deciding to reduce speed the observer should:

(a) Make a substantial reduction so that the action will be noticed by other vessels using radar.
(b) Know to what speed he is, in fact, reducing.
(c) Know how long it will take to attain the new speed.

In the following explanation of the method used to determine the effect of reducing speed, it is again assumed that the target vessel was passing an imaginary buoy when it was first seen at 1200 hours.

Fig. 41 shows the relative positions at 1212 hours of the imaginary buoy and target vessel. They are marked W^1 and A^1 respectively.

If the observer has been steaming 330° T. at 10 knots, the position of the imaginary buoy at 1200 hours must have been at O (2 miles, 330° T. from W^1). The apparent motion of the target vessel from O to A^1 indicates a risk of collision.

Suppose, however, that the observer has only been steaming at

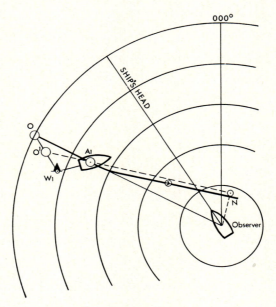

Fig. 41

5 knots. In this case the original position of the buoy would have been at O^1 (1 mile, 330° T. from W^1) and the target's apparent motion from O^1 to A^1, giving a nearest approach of 1·7 miles at N.

It will be appreciated from this that, if the observer reduced his speed instantly from 10 knots to 5 knots when the target was at A^1, the target's apparent motion would change from OA^1 to A^1N. As some time must elapse before the observer's speed could fall to 5 knots, the nearest approach would be actually less than 1·7 miles.

In this example a reduction to 5 knots would be unsatisfactory because both vessels would probably hear each other's fog signals and be required to stop their engines.

Summary and memory aids for the relative plot

In the simple plot the three sides represent in effect:

W^1O	What **O**bserver does in the plotting interval.
W^1A^1	What **A** does in the plotting interval.
OA^1	**O**bserver's and **A**'s movement combined giving the target's apparent movement.

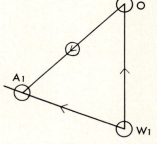

Fig. 42

When the observer takes avoiding action the target's new apparent motion is found by simply altering W^1O a corresponding amount to W^1O^1, O^1A^1 being the required new apparent motion.

Examples:

(1) Altering course 60° to starboard
Rotate W^1O 60° to W^1O^1.

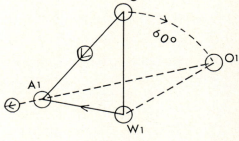

Fig. 43

55

(2) Reducing from 10 knots to 4 knots.
Reduce W^1O proportionately to the reduction in speed i.e. $W^1O^1 = 4/10\ W^1O$.

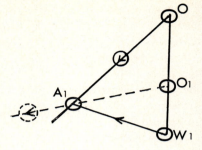

Fig. 44

(3) Reducing to half speed and altering 60° to starboard.
W^1O^1 = half W^1O turned 60° to Starboard.

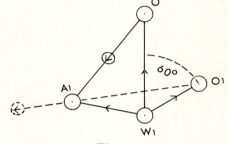

Fig. 45

(4) Increasing speed from 5 to 10 knots.
W^1O^1 is double W^1O.

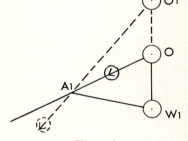

Fig. 46

In all the above cases, the new apparent motion line would have to be transferred to pass through the position of the target when the manœuvre was completed.

The true plot

In the true plot, the observer's position is moved across the

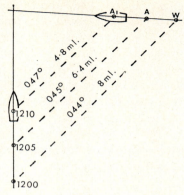

Fig. 47

plotting sheet as it would be on a chart and for this reason some may find it easier to understand. By laying off a target's range and bearing from these positions the target's actual course and distance made good in the interval is recorded.

Fig. 47 shows three positions of the observer at 1200, 1205 and 1210. The target's range and bearing from each of these positions

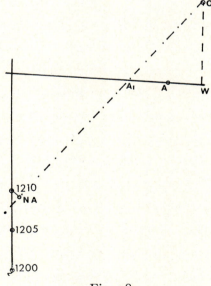

Fig. 48

57

was 044° T., 8 miles; 045° T., 6·4 miles and 047°, 4·8 miles respectively. If the first recorded position is named W and subsequent positions A, A^1, A^2, etc., then W to A^1 will indicate what target A has done.

To find the target's nearest approach from the true plot, it is necessary in effect to do the relative plot in reverse as in fig. 48. The distance and direction the observer steamed from 1200 to 1210 is laid off from W to O and a line from O through A^1 will be the target's apparent motion line, the distance from O to A^1 being the relative movement in 10 minutes. The nearest point on this line to the observer's position at 1210 will be the target's nearest approach.

The relative movement of the target which would result from the observer altering course or speed may be found by laying off from W to O^1 what he would expect to do in 10 minutes on the new course or speed.

Fig. 49, for example, shows the new apparent motion line (O^1A^1) which would result from the observer making an alteration of 60° to starboard and maintaining his speed. The nearest approach of the target shown in the figure would, strictly, only apply to an instantaneous alteration at 1210.

Although the true plot will give a quicker indication of the target's course and speed, it becomes more complicated than the relative

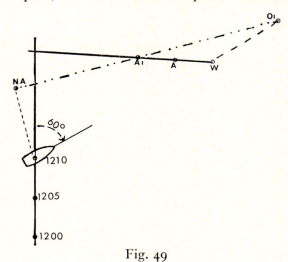

Fig. 49

plot when the target's nearest approach is also required. In considering the respective merits of the two methods of plotting some thought must be given to the fact that, although 5° or 10° error in finding the target's course would have little or no effect on the observer's subsequent actions, the same error in the apparent motion line could completely alter the situation particularly at the longer ranges (see page 34). Also, if in an emergency it is not possible to do anything but watch the radar screen, familiarity with the relative plot will enable the observer to have a greater appreciation of what is happening on a relative display.

A true plot can sometimes be made on a large scale chart, the target's range and bearing being laid off from the observer's positions found by shore fixes. This will give the target's course and speed over the ground but, when there is a strong tide, a completely wrong impression of the target's aspect may be obtained. The effect is similar to that on the true motion display when the course correction controls are being used. See example on page 15.

Radar plotting examples

Note: As bearings taken from the radar screen will be in 360° notation whether using 'Ship's Head Up' or 'North Up' displays, this notation will be used in the following examples: i.e.

$$300°R = 300° \text{ Relative } (60° \text{ on the port bow})$$
$$300°T = 300° \text{ True}$$
$$300°C = 300° \text{ Compass}$$

Aspect is given in the report in Red and Green notation as this gives a positive indication of the side a target is presenting, and avoids confusion with bearings:

i.e. Targets Aspect Red 60°

1. Your vessel is steaming 190°T at 12 knots.

Time	Target	Bearing	Range	Target	Bearing	Range
1145	A	300°R	8·4 ml	B	050°R	10 ml
1150	A	300°R	6·4 ml	B	055°R	8 ml
1155	A	300°R	4·4 ml	B	063°R	6·1 ml

Find the Course, Speed, Nearest Approach (NA) and time of nearest approach (TNA) of the targets.

59

2. Your vessel is steering 100°T at 6 knots.

Time	Target	Bearing	Range	Target	Bearing	Range
1300	A	326°R	5 ml	B	035°R	8 ml
1310	A	298°R	6 ml	B	035°R	6 ml
1320	A	280°R	8 ml	B	035°R	4 ml

Find the Course, Speed, NA and TNA of the targets.

3. Your vessel is steering 210°T at 9 knots.

At 0330 *Target* A bore 030° Relative *Range* 12 miles
0335 ,, A ,, 030° ,, ,, 11 miles
0340 ,, A ,, 030° ,, ,, 10 miles

Find course and speed, NA and time of NA of target. What action would you take, if any, and when?

4. Your vessel is steering 210°T at 18 knots.

At 0400 *Target* B bore 330° Relative *Range* 9·5 miles
0407 ,, B ,, 333° ,, ,, 8 miles
0410 ,, B ,, 334° ,, ,, 7·5 miles

Find course, speed, NA and time of NA of B.

5. Your vessel is steering 090°T at 18 knots.

At 0920 *Target* A bore 097°T *Range* 9 miles
0925 ,, A ,, 095·5°T ,, 8 miles
0930 ,, A ,, 093·5°T ,, 7 miles

Make out a report for 0940.

6. Own ship is steering North at 12 knots.

At 1000 *Target* A bore Rt. ahead *Range* 10 miles
1005 ,, A ,, ,, ,, ,, 8 miles
1010 ,, A ,, ,, ,, ,, 6 miles
1015 ,, A ,, 349° Relative ,, 5·1 miles
1020 ,, A ,, 333° ,, ,, 4·5 miles

How many degrees has the apparent motion line altered?

How many degrees has the target altered course?

60

7. Your vessel is steering 000°T at 12 knots.

Time	Target	Bearing	Range
1000	B	330°R	10 ml
1005	B	330°R	8 ml
1010	B	330°R	6 ml
1013	B	326°R	5 ml
1018	B	304°R	3·4 ml
1023	B	261°R	2·8 ml

What action has been taken by the target?

8. Your vessel is steering 030°C at 15 knots. Time required to stop 12 mins.

Time	Target	Bearing	Range
1610	A	320°R	10 ml
1614	A	320°R	8·5 ml
1618	A	320°R	7 ml

If the head reach of your vessel when stopping is 1·5 miles, when should the engines be stopped so that the target will cross your heading marker at a minimum two miles range? Assume the target maintains its course and speed.

9. Your vessel is steering 180°T at 8 knots.

Time	Target	Bearing	Range
1124	A	Rt. Ahead	10 ml
1131·5	A	,, ,,	8·5 ml
1139	A	,, ,,	7 ml

If target A is a light vessel, find the set and rate of tide. What alteration would have to be made at 1139 to pass two miles to the eastward of the light vessel? (Assume instantaneous alteration.)

61

10. Your vessel is steaming North at a constant speed.

Time	Target	Bearing	Range
0915	A	009°T	4·5 ml
0920	A	008·5°T	3·6 ml
0925	A	008°T	2·7 ml

Target A is a light vessel and the current in the vicinity is known to be setting 045°T. Find the rate of drift and the speed of own ship.

11. Your vessel is steering 240°T at 18 knots.

Time	Target	Bearing	Range
2105	Y	300°R	10 ml
2110	Y	300°R	9 ml
2115	Y	300°R	8 ml

If your speed was reduced instantaneously to 9 knots at 2115, what would be the NA of 'Y'?

12. Your vessel is steering 180°T at 12 knots.

Time	Target	Bearing	Range
0710	X	020°R	8 ml
0715	X	020°R	7 ml
0720	X	020°R	6 ml

What would be the NA of the target if:

(a) Your speed is increased to 24 knots at 0720?
(b) Your speed is reduced to 4 knots at 0720?
(c) Your course was altered 60° to starboard at 0720?

13. Your vessel is steering 278°T, speed 17 knots.

Time	Target	Bearing	Range
1806	B	022°R	12 ml
1812	B	022°R	10 ml
1818	B	022°R	8 ml

At 1818 you alter course 40° to starboard. (Assume instantaneous alteration.) Make out a full report for anticipated circumstances at 1830.

14. Your vessel is steering 020°T at 18 knots.

Time	Target	Bearing	Range
2400	A	050°T	12 ml
0005	A	051°T	10·5 ml
0010	A	052°T	9 ml

At 0010 you alter course 60° to starboard and reduce to 9 knots. (Assume instantaneous alterations.) Make out a report for time of NA. What will be the target's aspect when its bearing is 020°T.

15. Your vessel is steaming 000°T at 6 knots.

Time	Target	Bearing	Range
1200	A	020°T	12 ml
1200	B	060°T	12 ml
1200	C	335°T	12 ml
1205	A	020°T	11 ml
1205	B	060°T	11 ml
1205	C	338°T	10·5 ml
1210	A	020°T	10 ml
1210	B	060°T	10 ml
1210	C	341°T	9 ml

At 1210 you alter course 35° to starboard and increase speed to 12 knots. Find the target's NA and the change in target B's apparent motion. (Assume instantaneous alteration.)

16. Your vessel is steering 021°T at 10 knots.

Time	Target	Bearing	Range
0036	B	091°T	8·2 ml
0042	B	091°T	8·2 ml
0048	B	091°T	8·2 ml

At 0048 you alter course to 081°T and reduce speed to 5 knots. Make out a full report for 0100 and give the target's range when its relative bearing is 335° to your new course. (Assume instantaneous alteration.)

17. Your vessel is steering North at 15 knots.

At 0800 *Target* A bore 060°T   *Range* 9 miles
    0806   ,,   A   ,,   ,,   ,,   7·5 miles
    0812   ,,   A   ,,   ,,   ,,   6 miles

At 0812 you alter course to 060°T and reduce to 7·5 knots. Construct the target's new approach line and find the target's NA on this line.

18. Your vessel is steering 355°T at 15 knots.

At 1140 *Target* N bore 055°T   *Range* 8 miles
    1146   ,,   N   ,,   ,,   ,,   7 miles
    1152   ,,   N   ,,   ,,   ,,   6 miles

At 1152 you commence to alter course to 035°T and ring for ½ speed. At 1158 your vessel is steady on the new course and speed is 7·5 knots. Target now bearing 014°R, range 5 miles. Make out a report of the anticipated circumstances at 1204.

19. Your vessel is steering 180°T at 6 knots.

At 0200 *Target* Q bore Rt. ahead   *Range* 5 miles
    0210   ,,   Q   ,,   ,,   ,,   ,,   4·5 miles
    0220   ,,   Q   ,,   ,,   ,,   ,,   4·0 miles

At 0230 you commence to alter course to 215°T and ring for 12 knots.
At 0235 you are heading 215°T. Speed 12 knots. Target now bearing 320°R, range 3·1 miles. Make out a report for 0240.

20. Your vessel is heading 270°T at 5 knots.

At 1700 *Target* Z bore 025°R   *Range* 8·4 miles
    1718   ,,   Z   ,,   ,,   ,,   7·4 miles
    1736   ,,   Z   ,,   ,,   ,,   6·4 miles

At 1742 you commence to alter course to 320°T and maintain a speed of 5 knots.
At 1754 steady on new course 320°T. Target now bearing 331°R, range 5·4 miles.
Make out a report for 1812.

64

21. Your vessel is steering 240°T at 6 knots.

    At 1710 *Target* A bore 210°T   *Range* 8 miles
        1720   ,,   A ,,   ,,   ,,   6·75 miles
        1730   ,,   A ,,   ,,   ,,   5·4 miles

At 1730 orders are given to alter course to 310°T.
At 1740 you are steady on the new course.
Target now bearing 207°T, range 4·5 miles.
Make out a report for 1745.

22. Your vessel is steering 125°T. Speed 10 knots.

    At 0730 *Target* A bore 175°T   *Range* 12 miles
        0736   ,,   A ,,   ,,   ,,   11 miles
        0742   ,,   A ,,   ,,   ,,   10 miles

At 0748 you commence to alter course 30° to starboard and
ring for ½ speed.
At 0754 steady on 155°T. Speed 5 knots. Target now bearing
172·5°T. Range 8 miles. Make out a report for 0800.

23. Your vessel is stopped heading 075°T.

    At 1610 *Target* A bore 300°R   *Range* 5 miles
        1615   ,,   A ,, 300°R   ,,   4 miles
        1620   ,,   A ,, 300°R   ,,   3 miles

At 1620 orders are given to proceed 075°T at 12 knots.
    1625 steady on 075°T at 12 knots. Target now bearing
284° Relative, range 1·8 miles.

Make out a report for 1630.

24. Own ship steering North at 10 knots. Target A is a light
vessel. From observations below find the set and drift.

Time	Target	Bearing	Range	Target	Bearing	Range
1406	A	30°T	10 ml	B	343°T	10 ml
1412	A	30°T	9 ml	B	341°T	8 ml
1418	A	30°T	8 ml	B	337°T	6 ml

RW–E

What alteration of own ship to port at 1424 is required in order to pass 2 miles off this light vessel? Indicate the new apparent motion line of target B due to this alteration and give the NA. (Assume alterations to be instantaneous at 1424 and current constant for observation area.)

25. Observing ship is steering 300°T at a speed of 12 knots. Echo A is a light vessel. From the following radar observations, find the set and rate of drift experienced by the observing ship and the NA to the observing ship of target B.

Time	Echo	Bearing	Range	Echo	Bearing	Range
1305	A	325°T	10 ml	B	270°T	8 ml
1310	A	327°T	9 ml	B	266°T	7·2 ml
1315	A	329°T	8 ml	B	261°T	6·3 ml

Find the alterations of course to starboard required at 1325 to pass 2 miles off echo A and find the resultant nearest approaches of echo B due to these alterations. (Assume the alterations of course to be instantaneous at 1325 and that the effects of current remain similar over the area observed.)

26. Your vessel is steaming 000°T at 12 knots.

Time	Target	Bearing	Range
1800	A	003°T	11 ml
1805	A	004°T	9·5 ml
1810	A	005°T	8 ml

At 1810 you alter course to make the target's nearest approach on the port side 2 miles. Find the amount to alter assuming an instantaneous alteration and that the target keeps its course and speed.

Answers

1. A's Course 280°T. Speed 20·8 knots. NA Collision at 1206.
   B's Course 063°T. Speed 16·6 knots. NA 3·2 miles at 1207.

2. A's Course 001°T. Speed 15·4 knots. NA passed.
   B's Course 341°T. Speed 7·9 knots. NA Collision at 1340.

3. A's Course 107°T. Speed 6·2 knots. NA Collision at 0430.
   Action: Alter course 60° to starboard before target's range
   decreases to 5 miles.

4. A's Course 254°T. Speed 12·6 knots. NA 2·4 miles at 0444.

5. A's Course 058°T. Speed 7·5 knots. NA.
   Report: Target 3° on the port bow drawing aft.
   Range 5·0 miles decreasing. Aspect Red 157°.
   Nearest Approach 1·8 miles at 1003.

6. Target has altered course 90° to starboard.
   The apparent motion has only changed 45°.

7. The target has altered course 53° to starboard.

8. Stop the engines before 1623.

9. Tide is setting 180°T at 4 knots. Alter course to 155°T.

10. Tide is setting 045°T at 2·7 knots.
    Own ship's speed is 8·2 knots.

11. The nearest approach of Y would be 5·8 ml at 2146.

12. (a) NA = 1 ml at 0734.
    (b) NA = 3·1 ml at 0818.
    (c) NA = 4·6 ml at 0735.

13. B's Course 176°T. Speed 7·6 knots.
    Report: Target bearing 270°T drawing aft.
    Range 4·6 miles decreasing.
    Aspect Red 86°. NA 4 miles at 1835.

14. A's Course 302°T. Speed 7·5 knots.
    Report: Target bearing 008°T drawing aft.
    Range 6·5 miles increasing.
    NA 6·5 mls at 0034. Aspect Red 114°.
    Aspect when bearing 020°T Red 101°.

15. A's NA 2·8 at 1241.
    B's NA Collision at 1239.
    C's NA Collision at 1239.
    There will be no change in B's apparent motion direction but the rate of approach is increased.

16. B's Course 021°. Speed 10 knots.
    Report: Target bearing 079°T, drawing aft. Aspect Red 122°.
    Range 8·1 miles increasing. NA 8 miles at 0057.
    Range when bearing 335°R is 8·9 miles.

17. A's NA = 3·9 miles at 0825.

18. Course 314°T. Speed 13·2 knots.
    Report: Bearing 359°R, drawing aft. Range 4·3 miles, decreasing. NA 4 miles at 1210. Aspect Red 95°.

19. Q's Course 180°T. Speed 3 knots.
    Report: Target bearing 306°R, drawing aft. Aspect Green 161°. Range 2·7 miles decreasing. NA 2·4 ml at 0247.

20. Z's Course 235°T. Speed 2·4 knots.
    Report: Target bearing 314°R, drawing aft. Aspect Red 141°.
    Range 4·7 miles decreasing. NA 4·4 at 1828.

21. A's Course 341°T. Speed 4 knots.
    Report: Target bearing 252°R, drawing aft. Aspect Green 43°.
    Range 4·4 miles decreasing. NA 4·1 miles at 1808.

22. A's Course 060°T. Speed 8·5 knots.
    Report: Target bearing 013°R, drawing aft. Aspect Red 72°.
    Range 7·2 miles decreasing. NA 5 miles at 0831.

23. A's Course 195°T. Speed 12 knots.
    Report: Target bearing 220°R, drawing aft. Aspect Red 80°.
    Range 1·4 miles increasing. NA 1·3 miles at 1629.

24. Set 106°T at 5 knots. Course alteration 18° to port. NA 0·9 miles at 1435.

25. Set 025°T at 2·4 knots. Course alteration 7° or 51°. NA 5 miles for an alteration of 51°. Time of NA 1325. NA 4·2 miles for an alteration of 7°. Time of NA 1337.

26. Course alteration 33° to starboard.

six

Proceeding in reduced visibility

The master's first problem when proceeding in reduced visibility is to decide upon an appropriate speed. This is an extremely difficult task and unfortunately, when errors of judgement are made, they may never be fully appreciated unless a collision results.

The term 'moderate speed' cannot be defined to suit all circumstances and conditions and the master must base his judgement on variable factors which include:

(1) The range of visibility—an elusive factor which is hard to estimate, particularly at night, and one which may vary in a short space of time.

(2) The stopping distance of the vessel—this must vary according to the initial speed, wind, sea, and loaded condition.

(3) Knowledge of the locality—including the possibility of meeting ice, small craft, crossing traffic, tides and currents.

The great advantage of radar is not that vessels may proceed at faster speeds, but that they can avoid the dangers and frustrating delays inherent in a close-quarters' situation. Excessive speeds may mean that the advantage gained by the early detection of target vessels is lost: for instance, if two vessels meeting find it necessary to reduce their speeds immediately radar contact is made, the gradual reduction in their speeds as the vessels run their way off can make it impossible to ascertain with any accuracy their probable nearest approach. This could result in further action which might introduce a collision risk rather than avoid it.

The ideal speed when using radar might be said to be one which would allow:

(1) Time to determine the courses and speeds of targets in the vicinity. This would depend upon the probable density of traffic and the observer's ability.

(2) Time for the particular vessel to complete any necessary manœuvre to avoid collision.

(3) Time to establish that the action was satisfactory.

(4) Time to stop if the target was found to have re-introduced a collision risk.

These times together must obviously not exceed the total time available, which can be estimated roughly from the radar's normal detection range (or range scale in use, if less), and the probable maximum rate of a target's approach. All these times can be determined by practice and experience.

Bare Steerage Way

In many circumstances where it is necessary to stop the vessel, it is also important to avoid presenting the ship's beam to an approaching target or allow the ship to fall across the bows of a following vessel. In such cases it would be prudent to maintain enough way on the vessel to keep heading in the safest direction.

Scanty information

Navigators who manœuvre in fog when in possession of less information than they would consider essential in clear weather must, if they have the means of obtaining that information, be considered to be acting on 'Scanty information'. For instance, in clear weather, it is inconceivable that any officer would give instructions for an alteration of course or speed merely upon being told that another vessel was approaching on a steady bearing. He would first look to see which way that vessel and others in the vicinity were heading. It follows that the radar user, who has the facilities to do so, should ascertain the nearest approach and courses of other vessels before attempting to avoid a close-quarters' situation.

Close-quarters' situation

Before a close-quarters' situation can be avoided a pre-requisite is to decide what, in fact, is a close-quarters' situation. Two suggestions are:

(1) When the navigator's freedom of action would be restricted or lost, as when a fog signal of another vessel is heard forward of the beam.

(2) When an unexpected manœuvre by a target vessel could not

be ascertained with reasonable accuracy in sufficient time to avert a collision.

Early and substantial action

It was demonstrated in chapter four that a course alteration of not less than 60° is desirable if it is to show effectively on the radar screens of other vessels. This alteration should be made early to allow for further action should the target re-introduce a risk of collision.

Fig. 50 shows that an early alteration does not necessarily increase the distance steamed. Vessels A and B are meeting end on and it can be seen that the distance A must travel is the same if an altera-

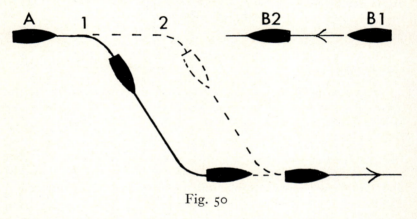

Fig. 50

tion of 60° to starboard is made at 1 or 2. An alteration at 1, however, has the decided advantage that B will be less likely to be tempted to alter course to port, and, even should B do so, there will be time for A to establish this fact and take other action.

It is advisable to take early action even when it is decided to stop. In fig. 51, vessels A and B are crossing at 90°. Now if vessel A stops when B is at 1, speed may be resumed when B arrives at 2. If A stops when B is at 3, then speed may be resumed when B is at 4. In both cases the time A will be stopped is the same, but in the second case it is possible that B will hear A's fog signal and stop, thus defeating A's object in avoiding a close-quarters' situation.

Note. Two prolonged blasts by A is not an invitation for B to proceed, irrespective of whether B has radar or not.

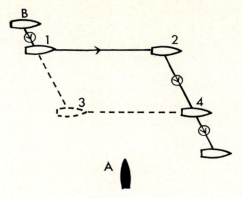

Fig. 51

Avoiding action based upon the information obtained from radar must often differ considerably from that taken in clear weather, but when a choice of solutions includes one which corresponds with clear weather practice, then this is usually preferred. In all cases due regard must be given to the possibility of sighting the target vessel and the need to apply the appropriate rule.

Targets on the Starboard Side

When a target is approaching on a collision course from ahead or on the starboard bow, navigators invariably prefer to make a starboard alteration of course as this follows normal procedure in clear weather.

Fig. 52 shows four possible cases.

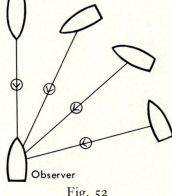

Fig. 52

It can readily be seen that in each case (fig. 52), a starboard alteration sufficient to put the target on the port bow (preferably with a minimum alteration of 60°) must succeed in avoiding a collision *as long as the target does not alter course to port.*

When a vessel is approaching on or near the starboard beam as indicated in fig. 53, then the large alteration required and the possibility of the target altering to port makes a starboard alteration less attractive and prudence suggests some alternative action. A round turn to port is not to be recommended, as the end result is the same as for a starboard alteration. Round turns are particularly dangerous when using a 'ship's head up' display as it is impossible to determine what action the target is taking whilst carrying out this manoeuvre. If a turn to port is decided upon then the vessel should be steadied with the target astern or on the port quarter until it has been established that it is quite safe to continue the alteration.

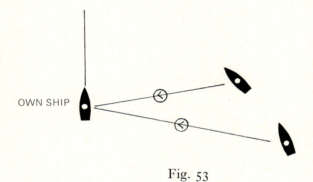

Fig. 53

An alternative solution to this situation, particularly where room is limited, is to make a large reduction in speed, always bearing in mind the time it will take to make this reduction.

When a target is closing on a steady bearing from more than two points abaft the starboard beam as in fig. 54, then it would be reasonable to assume that the target will take the necessary avoiding action. If, however, the target continues to close towards a close-quarters' situation, then the observer can alter course to put the target astern or on the port quarter, keeping the target in this position until it is safe to resume course. Where the target fails to take any action this may mean that the turn to port should be continued.

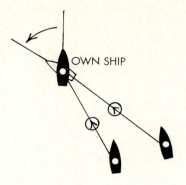

Fig. 54

It will be appreciated that in all the manœuvres to avoid a target on the starboard side which have been mentioned so far, the result would be an anti-clockwise change in the target's compass bearing.

Targets on the Port Side

When targets are approaching on the port side of the vessel, the navigator will find it more difficult to decide upon the most appropriate action. One reason for this is that he is unable to follow clear weather procedure and keep his course and speed; another reason is that it can be much more difficult to take action which will result in an anti-clockwise change in the target's compass bearing and so complement the target's probable action.

It is, of course, important that any action taken should not negate that action which will most likely be taken by the target and it is probably worth considering this in more detail.

Targets on the Port Bow

Figs. 55 'a' and 'b' (overleaf) each show a target closing from 30° on the port bow and indicate the change in the target's apparent motion which would result from an alteration of 60° to starboard by 'own ship'. Fig. 'a' shows the target proceeding at the same speed as 'own ship' and fig. 'b' a target proceeding at twice 'own ship's' speed. In each case it can be seen that:

(1) The target's compass bearing will change anti-clockwise.
(2) There is a reduction in the rate of approach.
(3) The action would be complementary to a starboard alteration of the course by the target.

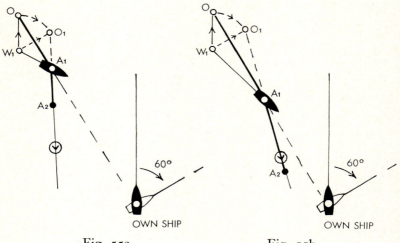

Fig. 55a   Fig. 55b

An alteration of course to port would be unwise because of the strong probability of the target altering course to starboard. A reduction of speed in this case may not give an adequate clearance if the target maintains its course and speed, particularly if 'own ship' carries her way for a long time.

Targets broad on the Port Bow

Figs. 56a and 56b, show targets approaching from 60° on the port bow, 56a indicating a vessel proceeding at the same speed as 'own ship', and 56b a target proceeding at twice 'own ship's' speed.

In these cases, whilst an alteration of 60 degrees to starboard by 'own ship' would not negate the effect of a starboard alteration of course by the target, such action alone would not resolve the situation. Fig. 56a shows that an alteration of 60° to starboard would put 'own ship' on the same course as the target, whilst in fig. 56b, the target continues to approach along the same apparent motion line. However, the reduction in the rate of approach would allow more time for the target to take action and put 'own ship' in a favourable position to make a further alteration of course to starboard if necessary.

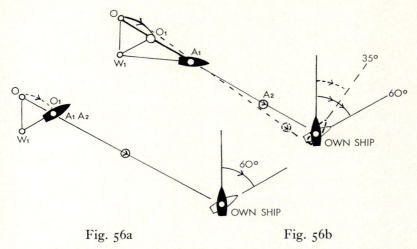

Fig. 56a   Fig. 56b

A smaller alteration by 'own ship' of some 35° to starboard to put the target just abaft the port beam, would produce an anti-clockwise change in the target's compass bearing and allow 'own ship' to cross ahead of the target, although this may be too close for comfort if the target is relatively fast. (See dotted apparent motion line in fig. 56b.)

Reducing speed is an alternative solution although the target may well do the same. Should the target alter course to starboard it would be reasonable to assume that it was aware of 'own ship's' position and that it would pass clear.

Again a port alteration would be dangerous if the target altered to starboard.

Target on the Port Beam

A target which is closing on the port beam must be a faster vessel, but this does not mean that the rate of approach will be fast, and in this situation there is probably no urgency to take avoiding action. If it becomes apparent that the target may not take action an alteration away from the target can be made. It will be noticed that apart from increasing speed *'own ship' cannot, by his action alone, make the bearing of the target change in an anti-clockwise direction.*

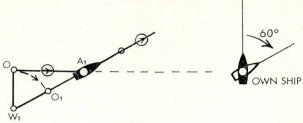

Fig. 57

Fig. 57 shows the effect of an alteration of 60° to starboard by own ship:

(1) The target's compass bearing will change clockwise.
(2) There will be a reduction in the rate of approach.
(3) The alteration would not negate the effect of an alteration to starboard by the target.

A reduction of speed is an alternative solution, although this would probably have to be made earlier than an alteration away from the target.

Target on the Port Quarter

Here, as for vessels on the starboard quarter, it is reasonable to expect the target to take the necessary action. If this is not done in reasonable time then the target could be placed astern.

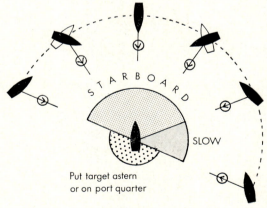

Fig. 58

The Steady Bearing

Fig. 58 shows the general pattern of possible action that may be taken when a target is on a steady bearing.

Bearing changing anti-clockwise

Fig. 59 shows that similar action may be adopted when there is a slight anti-clockwise change in the target's compass bearing.

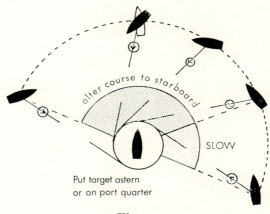

Fig. 59

Bearing changing clockwise

Situations in which two vessels are going to pass clear but too close can be more difficult to solve than actual collision cases.

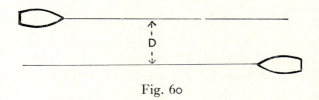

Fig. 60

If, in fig. 60 the passing distance 'D' between the two vessels is very small it is most probable that, when using radar, they would both alter course to starboard. As 'D' increases, however, so will the observer's doubts about an alteration of course to starboard. Now, if one vessel does alter course to starboard and the other vessel decides to increase the passing distance by altering course to port, the result could be disaster.

When the observer is undecided whether or not to alter course to starboard, it would probably be best to reduce speed to bare steerage way. A port alteration should only be attempted if it is reasonably certain that a collision could not result if the target alters course to starboard. This would probably mean putting the target abaft the beam until resuming course would give a satisfactory clearance. Similar bold action would be required if an alteration to starboard is made, a careful watch being kept on the target to see that the action is effective.

A similar problem to this occurs in clear weather when two vessels are meeting nearly end on but green to green. In this case, however, if one of the vessels decides to alter course a small amount to starboard so as to put the other vessel on his port bow, his action can be seen immediately, and both vessels would then accept a red to red passing. This simple solution to what has proved a most hazardous exercise when navigating by radar calls for some substitute for side lights during reduced visibility. (See Radar Beacons on page 95.)

Where the general pattern of collision avoidance is to favour an anti-clockwise change in a target's compass bearing, it is not really surprising that, when there is originally a slight clockwise change in bearing, the situation will be more difficult to solve and so potentially more dangerous.

Fig. 61 shows a target vessel which is proceeding at twice 'own ship's' speed and whose bearing changed some 3° as its range decreased from 7 to 5 miles. This change in bearing would give a nearest approach of about one mile. It can be seen that an alteration of 60° to starboard by 'own ship' could result in a collision. The big danger in this situation is that the unwary navigator may think that he has simply to alter to starboard to cross the target's *apparent motion line*, whereas, in fact, if the target does not alter course 'own ship' will have to cross the target's course line—in this case, an impossible task at these relative speeds. 'Own ship' is not only intercepting the target but presenting the full length of his vessel for the target to hit.

80

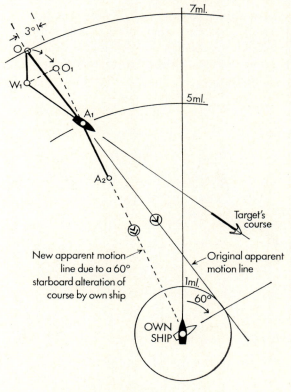

Fig. 61

Fig. 62 shows several cases of targets approaching with an initial clockwise change in bearing. Considering each case separately it is evident that whilst a large reduction in speed may be satisfactory for targets A, B, C and E it would be of doubtful value for targets D and F.

A large starboard alteration of course is a possible solution for target D and E but it may be best to put target F astern.

The reader may find it rewarding to consider separately each of the cases depicted in fig. 62 and make a note of the actions he would expect 'own ship' and the target to take. As the diagram actually shows only three complementary situations (Target B and 'own ship' is the same case as 'own ship' and target F with their positions reversed), it should prove of interest to pick out the other complementary pairs and see if the actions decided in each case are consistent.

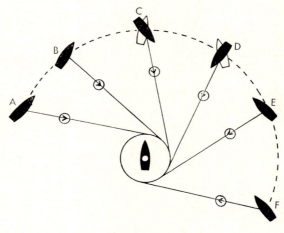

Fig. 62

Acceptable passing distances

When there is a slight clockwise change in a target's bearing, the passing distance at which individual navigators will act to introduce an anti-clockwise change in a target's compass bearing or accept the clockwise change will be influenced by:

82

(a) the size of his vessel.
(b) the present visibility.
(c) the locality.
(d) the traffic density.
(e) the relative speeds of the vessels.
(f) recent experience. (Following a long ocean passage, the same navigator will probably act differently from the way he would act after a long period in heavy traffic.)

The many variable factors involved make it impractical to produce a simple rule which would indicate when action should or should not be taken. Indeed, if some arbitrary passing distance was introduced it could prove more hazardous than helpful; inaccuracies in measurement could make one navigator believe that he was inside and the other believe he was outside the accepted figure. Over-confident action based upon this information could result in a collision rather than the avoidance of one.

Where a border-line case arises, it is particularly important that any action taken should be bold and early so that it will be conspicuous on the other vessel's radar screen.

Before comprehensive rules can be introduced for vessels navigating by radar, there are several difficult problems to solve with regard to identification, communication and coping with vessels not using radar. If, for example, one vessel found it necessary to act contrary to a prescribed rule, then some means of indicating its intentions would be essential.

Routing

Unfortunately, the considerable distances involved in collision avoidance by radar, frequently mean that the navigator must cope with more than one target vessel, and the necessary miss distance may make it impractical to deal with them one at a time as may well have been the case in clear weather. It follows that when navigating in multi-ship situations considerable skill is required. One solution to the problems of congested traffic is routing, but this requires the co-operation of all concerned and its introduction, as was the case with radar, has brought problems of correct usage. The IMCO book on Ship's Routing provides an extremely useful guide.

An alteration of speed, either alone or in conjunction with an alteration of course, should be substantial.

In order that both the observer and a target vessel can make an early estimation of the effect of a reduction in speed, the reduction should not only be substantial but carried out as quickly as possible.

It is perhaps not generally appreciated that large vessels may take an hour or more to stop without stern power and, during that time, travel 6 or 7 miles (in the very large super-tankers this distance can be doubled.) Even an emergency stop can take 10 minutes or so. Such vessels may be well advised to use stern power before an emergency exists and when any temporary loss of control will not prove hazardous.

An alteration of speed in conjunction with a substantial alteration of course can have little merit in an end on situation since the object

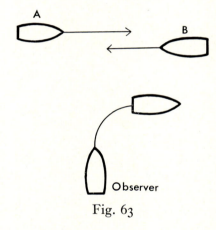

Fig. 63

of the course alteration is to get out of the target's path and a speed reduction would tend to defeat this.

When two targets are crossing, as in fig. 63, then a bold alteration to starboard, together with stopping the engines, may well be the best solution. This would avoid approaching the target's meeting point while running the way off the vessel and at the same time allow target A to overtake.

seven

Coastal navigation

In coastal waters, it is not always possible to avoid a close-quarters' situation and, in consequence, much of the advantage gained by the early detection of targets by radar is lost. Early warning of such a situation, however, does enable the navigator to begin taking the way off his vessel earlier than would otherwise be the case.

Alterations of course at close range are often unwise, due to the inability of the navigator to indicate the intended direction of the proposed alteration to the other vessel, and last minute attempts to use stern power can have unpredictable results. It is therefore desirable that full use should be made of the early warning of a close-quarters' situation to take the way off the vessel in good time. This means that the navigator should be aware of the stopping distance of his vessel at various speeds and under different loaded conditions, together with the effects of wind and sea. A table of stopping distances, as indicated on page 86, would prove of considerable value on all vessels.

The minimum range to which a target on a *steady* bearing should be allowed to approach before stopping can be found by multiplying the vessel's stopping distance by the decrease in the target's range while the observer steams 1 mile.

Example. If the observer's stopping distance is 2 miles and the target's range decreases from 7 to 5 miles whilst the observer steams 1 mile, then the engines should be stopped before the target's range falls to:

$$\text{(Stopping distance)} \times \text{(decrease in range)}$$
$$= 2 \times 2 = 4 \text{ miles}$$

Whilst it is appreciated that, in a congested situation, the navigator may not wish to indulge in mental arithmetic, consideration given in advance to this sort of problem could assist in determining what might be a reasonable speed under various conditions.

Table of stopping distances

Name of Vessel..............................
   Conditions during trial:
      Displacement............tons   Draught.........Frd..........Aft.
      Wind..........................   Force............   Sea............

| STOPPING DISTANCES ||||||||
| Manœuvring Speeds ||| Stopping with Stern Power || Stopping without Stern Power ||
Teleg.	Knots	Revs.	Distance	Time	Distance	Time
Full						
Half						
Slow						
D. Slow						

Normal Full Speed ....Knots ....Revs.
Notice required before Stopping ....

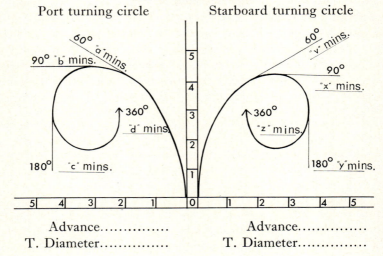

AT FULL SPEED
Port turning circle   Starboard turning circle

Advance...............   Advance...............
T. Diameter.............   T. Diameter.............

When it is possible to alter course to avoid close-quarters' situations, vessels with the land to port will probably, when meeting other vessels end on, alter course to starboard and move away from the shore, while vessels with the land to starboard will tend to move further inshore. This should always be borne in mind when coasting, the folly of always attempting to hug the coast being indicated in the following example.

Fig. 64 shows two vessels A and B meeting end on, close inshore. A is in the happy position that he has ample room to alter course to starboard, but B is not. In such circumstances, B may be tempted to alter course to port and should A alter to starboard, the risk of collision remains. The responsibility for this predicament arising must largely fall upon A, because B may well have been forced into

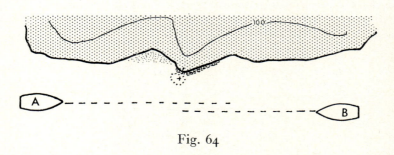

Fig. 64

this position by a series of end on situations. The moral here is, quite simply, that vessels with the land to port should always leave ample room for vessels which they meet end on to alter course to starboard. This, however, does not imply that vessels with the land to starboard should hug the coast. It must be recognized that ship-masters will vary considerably in what they consider to be a safe distance off shore. The danger here only exists for those who either like to keep close in shore or have no alternative.

A possible argument against vessels standing well off shore is that the distance en route would be considerably increased. It is, of course, true that some extra distance will have to be covered but possibly not as much as one may imagine. Fig. 65 illustrates that the increased distance required to pass 6 miles off a point instead of 3 miles off on a 40-mile run is negligible, about 3·5 cables, in fact.

Increasing the distance off shore, where there is a large alteration

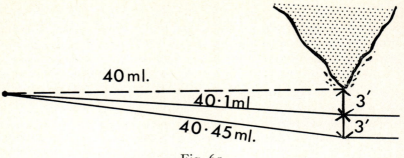

Fig. 65

round a headland, will add to the distance steamed but this is amply compensated for by the increased safety factor. It can be seen in fig. 66 that, by *doubling the distance off*, the manœuvring area inside the course line is *four times as great*, a valuable consideration when several vessels are rounding the headland in different directions.

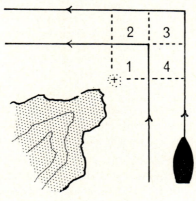

Fig. 66

Position fixing by radar

The identification of shore objects is a major difficulty in fixing the ship's position by radar. Lighthouses, although conspicuous to the eye, may return echoes of much less strength than nearby radio or electricity pylons and radar beacons are still in the experimental stage.

A radar range and bearing of a point of land, particularly at long range, cannot be regarded as a reliable fix for two reasons:

(1) The numerous errors to which radar bearings are subject.
(2) The possibility of mistaking high land inshore for the coastline.

The second reason shown in fig. 67 is of particular importance to fishing vessels using radar to keep outside restricted fishing areas.

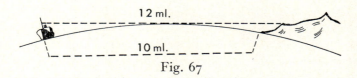

Fig. 67

Where a fix is obtained from two points of land, a double check using both ranges and bearings, will indicate if the points being used are correctly identified.

Fig. 68 shows how this would apply when using the extremities of an island. If point C, being low, was not shown on the radar, and high land at B was taken in error, the fix by two ranges would be inside the vessel's actual position and the fix by bearings outside.

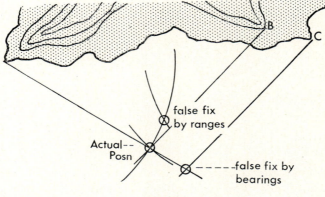

Fig. 68

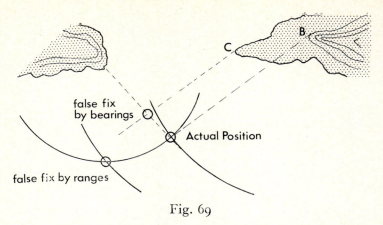

Fig. 69

Fig. 69 shows the effect when using the near points of two islands or land masses. Here it will be noticed that the false positions by bearings and ranges are reversed, the fix by ranges being outside the vessel's true position and the fix by bearings inside.

Where three positively identifiable targets are available, a fix by radar ranges should be reliable, providing the targets subtend fairly wide angles at the observer as in fig. 70.

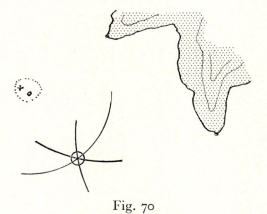

Fig. 70

Contour lines

On modern charts, places on shore having equal height are joined by contour lines. These are of considerable value to the radar observer as they give an indication of the type of echoes he may expect. It can be seen from fig. 71 that, where the contour lines are well spaced, the land will have an easy slope and may not provide good radar echoes. Where the contour lines are close together, the land will rise more rapidly and have better reflecting properties particularly when the lines are at right angles to the line of bearing.

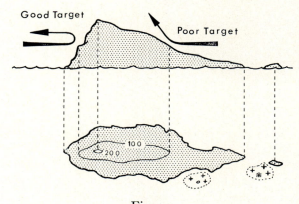

Fig. 71

Parallel index lines

Some mechanical bearing markers on the radar display have a series of lines scored on the perspex parallel to the bearing marker. These are called parallel index lines.

When a gyro-stabilized relative display is in use the parallel index lines provide a useful means of passing a required distance off a point of land or light vessel. The method is simply to set the bearing marker to the course it is required to make good, then manœuvre the vessel so that the target's echo moves down a parallel index line which passes the required distance off the centre of the display.

In fig. 72 (overleaf), the parallel index lines are set to make good a northerly course to pass a light vessel marking a shoal, to port. The difference between the direction of the heading marker and the parallel index line is the allowance which has had to be made for the westerly current.

On the ship's head up display, the practice of setting the parallel index lines parallel to the heading marker and using them to pass a given distance off a target can be dangerous.

91

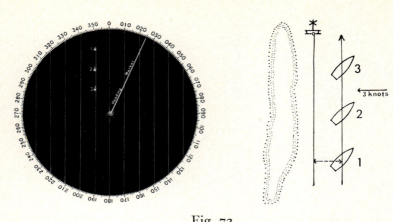

Fig. 72

If the vessel wishes to make a northerly course and to pass 2 miles off the light vessel which is marking a shoal, it can be seen that, by making a succession of starboard alterations of course, the light vessel can be kept on the parallel index line which passes 2 miles from the centre spot. Fig. 73 shows that this, however, does not stop the vessel going on to the shoal, a fact which may not be apparent when just gazing at the radar screen.

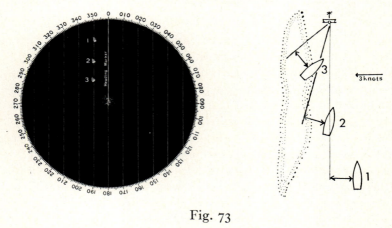

Fig. 73

Set and drift

The amount of set and drift can be found by making a simple plot of a known fixed object.

92

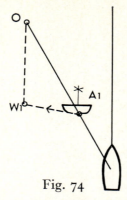

Fig. 74

Fig. 74 shows a light vessel on the observer's port bow, which was first detected at O. If there had been no tide the light vessel would have appeared later at W^1 instead of at A^1. Since the light vessel cannot move, the difference between these positions must be the amount the observer's vessel has been set towards the light vessel. The set and drift during the plotting interval will therefore be from A^1 to W^1.

Where it is required to find the amount to alter course to pass a given distance off a light vessel (see fig. 75 overleaf), allowing for the tide found by the relative plot, proceed as follows:

(1) Estimate the position that the light vessel will be in at the time of the alteration (A^2).
(2) From A^2 draw the required apparent motion line tangential to the circle of nearest approach. (N)
(3) Draw a line parallel to A^2–N through A^1.
(4) With centre W^1 rotate O to cut transferred apparent motion line at O^1.
(5) The angle O, W^1, O^1 is the amount that the course must be altered.

This procedure could theoretically be applied to finding the course alteration to pass a given distance off a moving target, but this is of doubtful practical value for four reasons:

(1) The alteration required may be too small to be detected by the target.
(2) The difficulty in estimating the target's position at the time of alteration.
(3) The assumption that 'own ship' alters course instantaneously.
(4) The target may also take action.

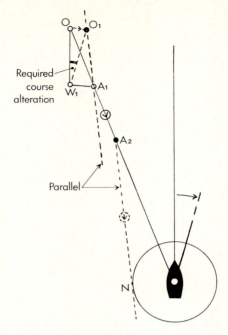

Fig. 75

Radar Beacons

Racon

This is a radar beacon which is switched on by the transmitted pulse from a ship's radar and it shows as a coded flash on the radar display at a range slightly exceeding that of the beacon, thus indicating the range and bearing of the beacon.

Ramark

This is a continuously transmitting beacon which sweeps the full radar band. It shows on the radar screen as a solid or dotted radial line on the radar screen indicating the bearing of the beacon.

As Ramarks do not have to be switched on by the interrogating vessel, they could be detected by a simple directional microwave receiver giving an audio signal suitable for homing or distress purposes.

Long Pulse Racon*

The Ramark has the disadvantage that at close range it can be received over a wide arc due to its continuous transmission, and as well as being ineffective as a directional aid it may obscure other targets. The Racon on the other hand will not show if the range of the beacon exceeds that of the radar range scale in use. It follows that neither beacon may be satisfactory as a form of leading light in inland waters when the radar is being used on a very short range. By using a Racon type beacon with a long pulse its signal may be seen as a second trace echo (i.e. after being switched on by one pulse it continues to transmit and is displayed as a radial line on the radar screen after the second pulse has been transmitted.)

Future developments

Whilst all improvements to radar sets which reduce the possibility of human error must be welcome and will be of value on some vessels irrespective of cost, the main objective must always be to provide aids which are effective, simple and economically viable for all types of craft.

It was suggested in chapter one that radar alone is little better than a blind man's stick, and this must remain true however sophisticated radar sets may become. However, it seems reasonable to suppose that in time radar beacons will be able to aid the navigator in reduced visibility in much the same way as lights do at night.

Radar Beacons (Ralights)

Colour television must have raised the question in all navigators' minds, 'When colour radar?' and no doubt this will arrive in due course, but meantime there is no reason why coded black and white signals should not go a long way towards giving the navigator much of the essential information required for safe radar navigation. Some indication of the possible use of radar beacons may be obtained by considering the type of lights they may effectively supplement. These could include:

---

* Some experimental work on this type of beacon is being done at the Hull Nautical College.

Light vessels
Leading and sector lights
Buoy and docking lights
Identification signals
Ships' Lights.

Light Vessels

It is now accepted that Racons on light vessels can be of great assistance both for making a landfall and for identifying the light vessel amongst congested traffic. It might be argued, therefore, that their use should be restricted to this purpose to avoid possible confusion, but this is not to say that other uses should not be considered.

Leading Lights

The use of radar beacons for leading lights could be most valuable in clear as well as in poor visibility, particularly in view of the ever increasing size of vessels using ports with restricted depths of water.

Buoys and Docking Lights

There is an undoubted need in some ports for the immediate recognition of buoys, dock entrances, riverside berths and navigational hazards. Transit marks for speed assessment and for checking radar alignment are also desirable.

Identification Lights

It has already been mentioned that Racons are a most useful means of identifying light vessels, but the problems of identification are much wider than this. Much thought has gone into the possible introduction of new rules for collision avoidance, but it is doubtful if new rules can be really effective until vessels are able to identify and communicate with each other, particularly when it is necessary to take some action which is contrary to the prescribed rules.

Traffic surveillance, special types of vessel, hampered vessels, vessels in distress and navigational hazards are all valid reasons for the use of identification signals.

Ship Lights

In many of the serious collisions which have occurred between vessels using radar, it is probably true to say that if each vessel had been able to determine which side light the other vessel was presenting, these collisions would have been avoided. There is a case, therefore, for ships to carry Racon type beacons which would produce different coded signals on the port and starboard sides with a combined signal for right ahead. (This could probably be achieved using low power beacons having twin aerials and a dividing screen.)

The many possible uses of radar beacons raises the obvious question of congestion of the radar screen and calls for some international agreements governing their use. At the same time it should not be forgotten that some types of memory radar displays have equally as much congestion and this is apparently no serious problem. One method of reducing congestion would be to have a special frequency fixed for beacons; this would allow the navigator to use the normal frequency to detect unexpected dangers and another frequency to obtain specific information about those targets. There is an obvious advantage in being able to read coded signals on a screen uncluttered by land, sea and rain echoes particularly if before these signals fade one can relate them to the normal picture at the flick of a switch.

Special beacon frequencies should require only slight modifications to existing radar sets, and provided the frequencies used were close together bearing errors introduced by slotted waveguide scanner squint would not be serious. Some experimental work has been carried out by Admiralty Surface Weapons Establishment, Portsmouth, in this field.

Perhaps a combination of fixed frequency beacons and colour television will be the answer for colour radar—it would be very nice to see if a target was showing a red or green response.

Communication

In addition to the need for means of identifying radar targets, an effective means of communication which can be readily understood by vessels of all nations is an urgent requirement. An obvious choice is VHF with an accepted international language; although there may be problems of air congestion, experience on crowded fishing grounds, where skippers are continually discussing their catches and agreeing to various manœuvres, would seem to indicate that it is not as serious as one might expect.

eight

Radar on small craft

There can be no doubt that the small low-power radar set is an extremely valuable aid for position fixing on the smaller vessel, particularly for those most subject to the vagaries of wind and tide, but perhaps a more cautious view should be taken of its use for collision avoidance in poor visibility. Bearing in mind the number of collisions which have occurred between vessels using radar, navigated by qualified officers with considerable radar experience, it is evident that the use of radar for collision avoidance by untrained persons could be very dangerous. This is particularly true when there are other limitations such as:

(1) Too few crew members to keep a continuous radar watch.
(2) Lively boat movement making plotting on paper impracticable.
(3) No facilities for plotting directly on the radar screen.
(4) Yaw making it difficult to take bearings.
(5) A low scanner giving a poor maximum range.
(6) Vibration and the possible ingress of water reducing the reliability of the radar.
(7) The possibility of not carrying anyone on board able to maintain the set.

Under these circumstances the most prudent use of radar may be to select a safe position well away from shipping and remain there until the visibility improves. Unfortunately, this is not always possible as the visibility may deteriorate when there is no alternative but to continue with the voyage. It is essential, therefore, to be prepared for this eventuality. This means that the navigator must know the capabilities and limitations of his set and be able to recognize maladjustments and low performance. He must also be aware of the difference between the real movement of targets and the apparent movement as seen on the radar screen as discussed in chapters 4 and 5. This is particularly important when plotting is not being carried out, as although one can formulate rules of thumb to help in interpreting radar information, such rules are not necessarily true in all cases. For example, it was mentioned on page 35 that when two vessels are approaching each other at the same speeds, the change in apparent motion when one vessel alters course will only show as half of the actual course alteration. Is this true in all cases? If one vessel altered course 360°, would the apparent motion only change 180°? Obviously not, the apparent movement must revert to its original direction; but where did the extra 180° come from?

Consider fig. 76. Here we see target A altering course in 90° steps, while the apparent motion only changes in steps of 45°. So when the target has *almost* turned 360°, the apparent direction has only changed 180°—but if the target completed the 360° turn the apparent movement would go back to its original direction. The extra 180° change occurred as the vessels altered from slightly converging to slightly diverging course—target B shows this clearly. When the target is a faster or slower vessel the change in apparent motion, due to an alteration from diverging to converging courses will be less than 180°.

Failure to appreciate this effect has probably contributed to several collisions where one vessel has altered course across the bows of a vessel astern.

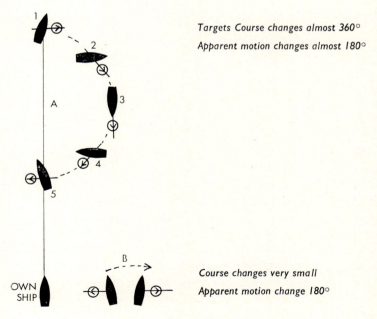

Fig. 76

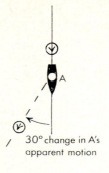

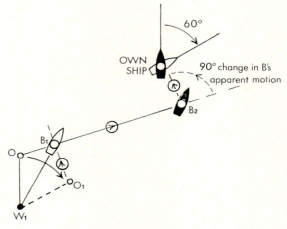

Fig. 77

Fig. 77 shows 'own ship' altering course 60° to starboard to avoid target A and it can be seen that the change in A's apparent motion is 30° as might be expected for a vessel of similar speed to 'own ship'. But the change in B's apparent motion is some 90°* which could result in a collision. This result is perhaps obvious from the figure, but a navigator who has not plotted the situation may fall into the trap of believing that he could alter course to starboard to run parallel with B's apparent motion line.

A closer study of this figure will show that knowing the target's course helps to decide not only what action to take but *when to take*

* The large change in apparent motion is due to 'own ship' altering from one side of B's course to the other.

it—which can be equally important. Compare a starboard alteration of 60° by 'own ship' when target B was at B^1 with the same alteration at B^2.

It is of little satisfaction, after such a collision, to know that the vessel astern could be censured for passing so close.

Having mastered the problems involved in interpreting correctly the information displayed on the radar screen, the navigator should give considerable thought to the various manœuvres discussed in chapter 6. It is important to remember that, although the actions may seem extreme for a small vessel, other vessels in the vicinity may not be small and a thoughtlessly made close approach may have serious repercussions.

When possible it is advisable for small craft to avoid channels used by very large vessels because once committed, a deep laden vessel in shallow water may

(1) be unable to alter course.
(2) have difficulty in steering.
(3) have to cross to the wrong side of the channel to obtain an adequate depth of water.
(4) be unable to stop.
(5) require the full width of a narrow channel, particularly at bends.

When the use of such a channel by a small vessel is unavoidable, it is important to keep to the correct side always bearing in mind the points 1 to 5 mentioned above.

nine

Clear weather practice

In dealing with the various aspects of radar navigation, reference has frequently been made to the need for practice in clear weather. To be really effective, however, practice must be properly organized and not left to the whim of the individual. Co-ordinated practice will ensure that all officers become familiar with the terms used in radar reporting and so avoid confusion in times of stress. In addition, the most suitable methods and facilities for radar plotting to suit the particular vessel will suggest themselves.

While the details of radar watchkeeping must depend upon the number of officers available, the skills required by each observer are the same and can be summarized under three headings:

(1) The operation of the radar set.
(2) Position fixing by radar.
(3) Collision avoidance by radar.

The operation of the radar set

Since all information obtained from the radar set must initially depend upon its being correctly adjusted, the master must ensure that all his officers:

(a) Know the function and position of all controls so that they can be correctly adjusted at night.
(b) Are able to recognize the indications of peak performance such as the extent of sea-clutter which may be expected under various sea conditions and the length of the performance-monitor plume.
(c) Immediately recognize maladjustments and their cause.
(d) Understand any idiosyncrasies that the set may have.
(e) Discover any personal tendency to make particular errors.

In order to gain and maintain proficiency in operating the radar set, particularly in areas where it may not actually be required for long periods, a weekly exercise could be introduced. The set should be switched on for a period which would allow each officer to practice and possibly correct maladjustments deliberately introduced by the master. A log entry could also be made regarding the extent of the performance monitor-plume.

Position fixing by radar

The convenience of radar for taking bearings and its advantage over the eye in measuring ranges has led to its wide use for fixing the vessel's position in clear weather as well as in fog. Even so, the navigator cannot really appreciate how much he may rely upon the accuracy of fixes obtained entirely by radar unless simultaneous checks on the vessel's position are made by other means. One way in which a regular comparison might be achieved is for the common procedure of several officers obtaining the noon position in open waters to be continued on the coast, each officer finding his position by different methods where possible.

On some coasts although there may be many objects from which a visual fix can be obtained, such as lighthouses, church spires, chimneys and beacons, none of these may show on the radar screen. It is therefore important that the navigator, when in areas subject to reduced visibility, should identify and mark on his chart suitable targets for future use. The type of targets required are those which:

(a) Return strong echoes. (This can be determined by a temporary reduction in the gain.)
(b) Are easily identified.
(c) Can be detected from all positions to seaward; i.e. are not in the shadow of high land or other obstructions.

Collision avoidance by radar

Regular clear weather practice in collision avoidance by radar is not always possible due to the lack of suitable targets, and consequently, full use must be made of opportunities as they arise. Actual manœuvres must, of course, confirm to the collision regulations and it is primarily due to this that radar simulator courses have been introduced. On these courses the navigator is able to manœuvre his vessel through more awkward situations in one week than he would probably meet in several years at sea.

What can and must be practised at sea, irrespective of subsequent avoiding action, is the skill with which a situation depicted on the radar screen can be correctly assessed. A test of proficiency in this would be for the master to take over the bridge and compare what he can actually see with a description given by his officers from radar information alone.

Whatever type of radar plot may be used for this, it is the measurement and recording of the target's position that takes the most time, particularly when the 'ship's head up' type of display is used. The observer should therefore concentrate initially upon attaining speed and accuracy in measuring a target's range and bearing.

When a target vessel is passing clear, an excellent exercise is to determine its nearest approach on a relative plot (see page 31) as this provides a means of checking the accuracy with which the target's relative positions have been recorded. Considerable satisfaction and confidence will be gained by the observer who, having set the variable range ring and bearing marker to the target's estimated nearest approach position, subsequently sees its echo pass through this point.

This type of exercise is suitable when only one officer is on the bridge as all the work involved can be completed when the targets are still at a considerable distance.

Finding the target's course and speed, as indicated on pages 42 and 43, will present no problems once the observer finds he is able to estimate the target's nearest approach with reasonable accuracy.

In situations where the target vessel must take avoiding action, a comparison between the situation as seen visually and as seen on the radar screen can be most rewarding. Particular points to note are:

(a) The effect of small alterations in the target's course on its apparent movement on the radar screen.

(b) The ranges at which vessels alter course in clear weather.

(c) The time which elapses before a target's alteration of course becomes obvious on the radar screen and the new course can be determined.

When it is the observer's duty to take avoiding action and the master or another officer is on the bridge ready to countermand any wrong order, alteration and resumption of course by radar so as to pass a minimum distance off a target vessel is not just a good exercise but essential preparation for manœuvres in fog.

ten

Plotting aids

R.A.S. (Radio Advisory Service) plotter

It is desirable, when plotting, that the bearings of targets should be recorded relative to compass north rather than the ship's head as this will avoid discontinuity in the echoes' movement when the observer alters course. To eliminate the necessity of converting relative bearings to compass, and vice versa, aids such as the R.A.S. plotter have been introduced. These have two bearing scales, one of which can be rotated and re-aligned each time the vessel alters course.

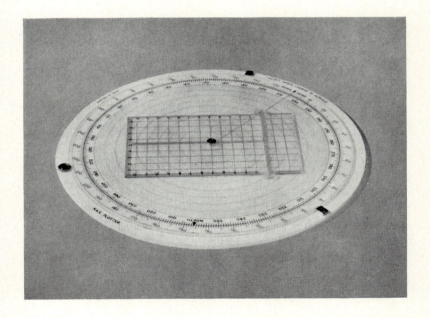

Track plotter

This does away with the necessity of using parallel rulers, dividers and protractors or compass roses and it can be used for either making a true or relative plot on plain paper. A fitted low voltage

light over the graduation pointer permits its use without other lights at night.

Plotting by perspex or acetate

This is only suitable for large radar screens not fitted with a magnifying glass but it provides an extremely simple and quick method of plotting and will be described in detail.

The equipment required consists of a blue chinagraph pencil and some thin perspex. The perspex is cut into discs having a radius equal to the 10-mile range ring on a 12-mile range (or 8-mile range ring on the 10-mile range).

To find the target's course and speed

After drawing a radial line on the perspex with the chinagraph pencil, place the perspex on the screen so that the line lies over the heading marker as in fig. 78. Mark $-x$ over the centre of the screen, draw small circles round each target and write down the time beside $-x$.

Keeping the perspex in the same position, make further marks

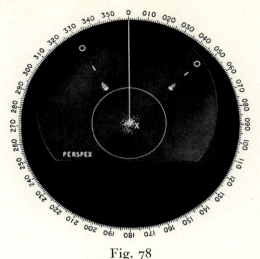

Fig. 78

over the target after steaming for 1 and 2 miles or other selected plotting intervals. If the circle and the two marks (o – –) are in line and equal distances apart, it is reasonable to assume that the target has kept a steady course and speed. Should the apparent motion of the targets indicate that they will not pass anywhere near the centre of the screen, then it is only necessary to continue marking the targets' positions until they are passed and clear. If on the other hand there appears to be a risk of collision, it will be necessary to find all the targets' courses before taking avoiding action. This is done quite simply by sliding the perspex down the screen the distance that the observer has steamed since the original circles were made. The distance being found by setting the variable range ring to the distance steamed and moving – x down to lie over the range ring as in fig. 79.

The circles on the perspex will now indicate the positions the targets would have been in had they been stationary (W^1 on the simple plot, fig. 80), and a line from these circles to the targets' present positions will be the course and distance made good by the targets in the plotting interval.

The movement of the perspex down the screen on the 12-mile range is limited to about a 5-mile run by the observer but, by this time, the targets will either be passing clear or it will be time to

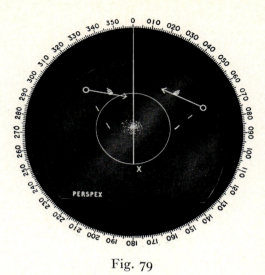

Fig. 79

tuun to a lower range scale. Should it be necessary to reduce range, a new piece of perspex should be used, the original with the targets' courses being placed handy for future reference if necessary. The new piece of perspex will be used in a similar manner to the first but a shorter time interval may now be used, giving a quicker indication of danger.

This method of plotting obviates the need to measure the ranges and bearings of targets and of drawing a vector diagram for each target. The time saved and the elimination of construction lines

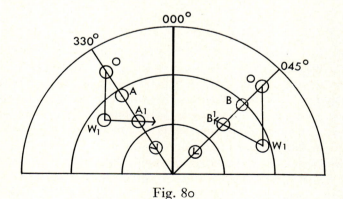

Fig. 80

means that the observer can cope with a greater number of targets. Plotting by perspex is equally suitable for North up or Ship's head up presentations.

Reasonable care must be taken when using this method to avoid errors due to parallax, although it would seem from exercises on a radar simulator that this is not a serious problem. In point of fact it has been found in simultaneous exercises, using perspex on one display and a plotting diagram on the other, that when a serious discrepancy has arisen, it is invariably owing to an error in reading or laying-off bearings on the conventional plot.

Position identification

It is possible when approaching land after several days in fog that there may be some difficulty in recognizing the coast configuration showing on the screen. Identification may be simplified by tracing the echoes on perspex and comparing them with the chart, due allowance being made for the difference in scale. If radial lines are drawn as indicated in fig. 81, a rough form of station pointer is obtained.

An outline of the coast or river banks may also prove useful to keep a check on the vessel's position when at anchor, particularly if there are no conspicuous objects available.

Vessels having to negotiate awkward places fairly regularly may

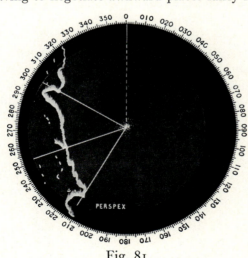

Fig. 81

also find that a tracing on perspex of the area in clear weather can be of future value in fog, particularly in identifying buoys and light vessels amongst vessels and boats at anchor. In this respect, one looks forward to the day when radar engineers will be able to supply radar screens with a large flat working surface. These would allow the navigator to construct and use his own perspex chartlets.

Perspex or tracing paper can also help to speed up the constructional work on an ordinary plotting diagram and may be used on the Kelvin Hughes photo display and the Decca automatic plot.

To find the target's course
(1) After marking the three positions of each target, O, A, A^1 and O, B, B^1, etc., on the plotting diagram, place a piece of perspex over it and draw a small circle round each target's original position. Mark the centre of the diagram X and a line indicating the ship's course as in fig. 82.
(2) Slide the perspex over the plotting sheet so that X moves the distance steamed in the plotting interval, but in the reverse direction. The circles made round each target will now correspond with W^1 on the simple plot and W^1 to A^1 will be the target's course and speed. (Fig. 83.)

To find the effect on the target's apparent motion if the observer takes action:

Stopping

When the observer's vessel is stopped the target's apparent motion will be the same as its actual course and speed. If the target's course is drawn on the perspex, the observer can see at a glance if stopping would be satisfactory. Due allowance can be made for the observer's stopping distance by sliding the perspex down the plot a corresponding amount.

Altering course

To investigate the effect of altering course, say 60°, to starboard.

After finding the target's course, as above, mark Y on the perspex over the centre of the plot and then slide the perspex so that Y moves along a line 60° to starboard of the heading marker, a distance equal to the distance steamed in the plotting interval. The original heading line on the perspex is kept parallel to the original course. From the new positions of the circles to A^1, B^1, etc., will

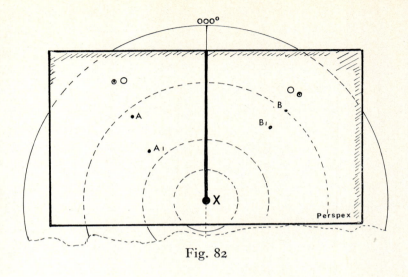

Fig. 82

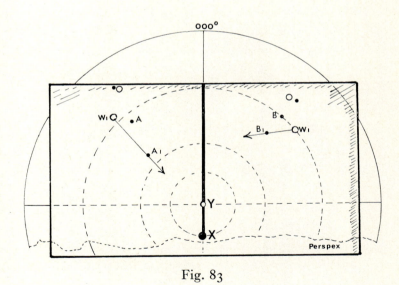

Fig. 83

111

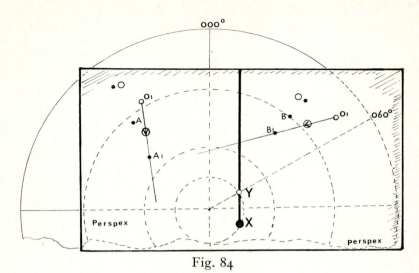

Fig. 84

be the new rate and direction of apparent motion, fig. 84, corresponding with O^1A^1 in the conventional plot.

The Decca marine radar system

The Decca Automatic Plotter when used in conjunction with the

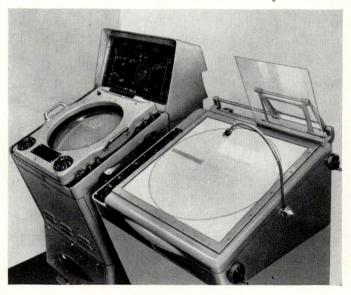

Decca True Motion Radar TM 969 provides an automatic means of recording the target vessel's position. The procedure is simply to set the bearing and range controls to the target's position and press a foot switch. This will automatically print the target's position, together with the time (if required), on a relative plotting diagram. A series of these positions will give the target's apparent motion line from which its nearest approach can be found.

An additional facility, called a predictor, enables the observer to determine the change in the target's apparent motion which would result from some contemplated action. The procedure being much the same as that mentioned in plotting by perspex.

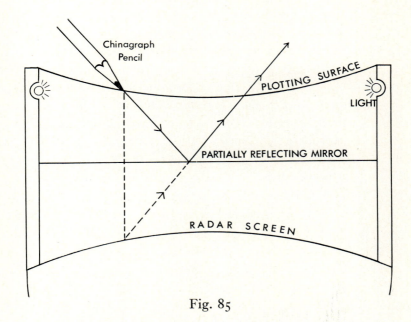

Fig. 85

The reflector plotter

This is a device for plotting directly on the radar screen which avoids possible errors due to parallax. It offers an ideal means of recording a target's position and, when there are not many targets, it is also possible to make simple plots. Parallax is eliminated by using a semi-reflecting mirror between the radar screen and the plotting surface, the latter having a curvature equal and opposite to the screen. This enables the reflection of a mark made on the plotter, directly above the echo, to coincide with the position of the target's echo on the screen from whatever angle it may be viewed. Chinagraph pencils are used and edge-illumination of the plotting screen makes the reflections and marks show clearly. When the illumination is switched off, the marks disappear and only the target echoes on the radar screen remain visible.

Kelvin Hughes photographic display

By employing rapidly processed photographs of a miniature radar screen, this equipment is able to project bright radar pictures on the under side of a transparent plotting table.

It is possible to have a new picture about every 3 seconds or, by using a longer exposure time, of, say 6 minutes, have a picture which indicates the movement of target echoes during that period.

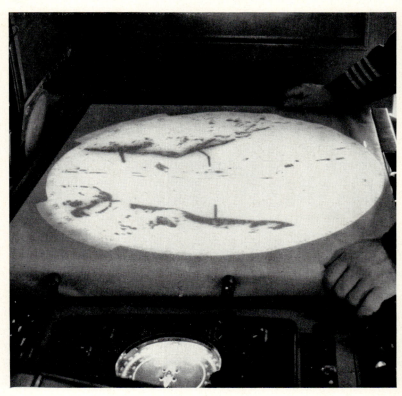

As the minimum exposure time (3 seconds) allows for one complete rotation of the radar scanner, the resulting picture is of even brightness and the targets are not subject to fading as on the normal radar screen.

Autaplot Ltd.

The Autaplot is a separate pedestal mounted device which can be used to record both true and relative plots simultaneously from an existing radar.

The true plot is made on a continuously moving sheet of transparent material over a relative plot made on a stationary sheet of red plastic. A warm stylus is used to plot selected ships, making a hole in the red plastic and a dot on the true plot. The holes representing the relative plot show white in daylight and green on an amber background at night.

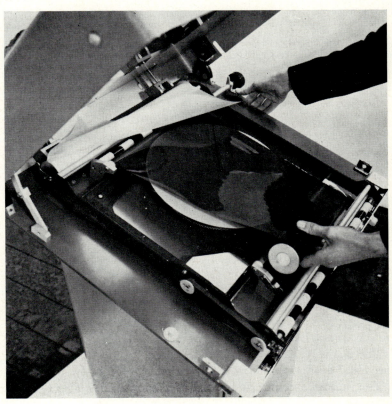

An optical device fitted to the radar hood is used to plot the required targets. By turning a radial line of light to the target's bearing and moving a range marker to its range, the position is recorded by pressing a thumb bar.

The machine is charged with enough material for some 2,000 miles of continuous plotting.

Compact (GEC–AEI) Radar

The Compact display console houses two 16″ C.R. tubes one providing a conventional chart plan radar picture in orange and the other showing automatic tracking markers in green. A special optical device is used to superimpose the two displays.

Early warning of an approaching target is provided by means of two guard rings at 11 and 9 miles round 'own ship'. When either of these rings is cut by an approaching target an audible alarm is sounded and a dotted marker indicates the bearing of the target.

Up to 12 and 11 circles called auto tracking markers can be placed round selected targets. Once placed these markers follow the targets and a computer analyses the information obtained. The course and speed of each target is indicated by a track ahead line, the length of the line being dependent upon the time scale set by the operator. An audible warning is given if any selected target's predicted movement will bring it inside a predetermined distance from 'own ship'. Positive identification of the target is given by the marker ring flashing on and off.

Compact can provide the operator with a trial change of course and speed when avoiding action is needed.

Digiplot

The Digiplot used in conjunction with a normal radar will analyse all radar echoes and display the 40 most threatening targets on a cathode ray tube. Presentation may be either North up or Ship's Head up, each target being displayed as a small circle with a vector line indicating its course and speed or relative motion. Land masses are shown as a dotted outline of the nearest edge.

There is an optional digital readout of the range, bearing, course, speed and nearest approach of any target.

The indicated course and speed of a known fixed target may be used as inputs to the set and drift controls to enable the plotter to show the course and speed of targets over the ground.

The Digiplot may be used for trial manœuvres and it will display the predicted situation after a trial course and speed has been inserted. Alarm controls may be set for a target's nearest approach of 0 to 5 miles with a warning time of 0 to 30 minutes.

Marconi Predictor Radar

The predictor stores radar information on a magnetic tape and displays it in true or relative mode as required. A novel feature of this radar is that when true motion is used, the observer's position remains at the centre of the screen, the tracks of 'own ship' and target vessel being indicated by a tail of three small echoes separated by the distance steamed by each vessel in two minutes. The picture is built up over four scan sweeps: the first scan shows a picture recorded six minutes ago and displayed round a centre point which is set back on the screen the distance steamed by 'own ship' in this interval; this is followed by the four minutes and two minutes ago pictures; the 'now' position shows 'own ship' at the centre again. This sequence is repeated every ten seconds.

As its name implies this set can be used to predict the result of a proposed action by 'own ship' before that action is taken.

APPENDIX A

MERCHANT SHIPPING NOTICE No. M.535

RADAR IN MERCHANT SHIPS: SITING PRECAUTIONS

Notice to Owners, Masters and Officers of Merchant Ships and Fishing Vessels

This Notice supersedes Notice No. M.411

GENERAL

This Notice, which replaces No. M.411 of August, 1957, is for information and guidance when radar is to be installed in merchant ships. For new ships, much of the advice is worth considering at the time when the ship is being designed, rather than at the stage of building.

It is the practice of the Board, after consultation with representatives of the users and manufacturers of radar equipment, to set down performance standards and a performance specification which, in their opinion, an efficient general purpose radar equipment should meet. These standards and specifications are subject to revision from time to time. The Board further undertakes to test a specimen model of a type of radar equipment and to compare its performance with the standards and the specification. Where the performance of the specimen model is found to comply with requirements, a certificate of type-testing is granted to the manufacturers.

In this Notice, the phrase 'type-tested set' is used to denote a set which has satisfactorily passed the Board's type-test and has received a certificate of type-testing.

PRECAUTIONS TO BE TAKEN BEFORE FITTING

1. Compasses

1.1. It is most important that units of radar sets that contain magnetic material should not be sited too close to magnetic compasses. Many units of radar sets contain enough magnetic material to make the minimum separation distances quoted in para. 7(a) of Notice No. M.417 (10 feet from a standard and 5 feet from a steering compass) insufficient. In accordance with the requirements of the Board's specification each unit of a type-tested set is tested by the Admiralty Compass Observatory to determine the minimum 'safe distances' at which it should be installed from both steering and standard magnetic compasses in order not to affect the accuracy of these compasses, and such safe distances are indicated on a tally plate on the unit concerned. A 'safe distance' takes account of both

119

the constant effect on a magnetic compass due to the presence of magnetic material and also any variable effect due, for example, to electrical circuits or to opening or closing of drawers. Thus, provided a unit is not placed in a position nearer the centre of the bowl of a magnetic compass than the prescribed safe distance, the unit may be installed or removed without any need for adjustment of that compass.

1.2. If, in installing a radar equipment in a particular ship, it should prove impracticable to site any particular unit of the equipment outside the tested safe distance as indicated on the unit concerned, a Board surveyor should be consulted. He will advise whether it is possible to site the unit concerned in a position nearer to the compasses than the indicated safe distance where the effect of such siting on the compasses will be stable and can be allowed for by compass adjustment.

1.3. Whenever a unit is fitted nearer a magnetic compass than the safe distance that compass should be adequately checked and, if necessary, adjusted. This should also be done whenever such a unit is removed, replaced or modified. Shipowners and masters are warned that these changes may seriously affect the compasses and disregard of this precaution may have dangerous consequences. It should also be noted that any maintenance that involves more extensive movement of parts of the set than can be brought about by finger pressure on the handles or catches may have an abnormal effect on the compasses while it is in progress.

1.4. Permanent structures such as masts that support units of the radar equipment housed above deck and huts that house equipment or parts of equipment will normally be considered as part of the ship's structure and will not be marked with 'safe distances'. After the installation of such items near a magnetic compass, the same precautions with regard to the accuracy of that compass should be taken as are taken after any other structural alteration.

1.5. In determining 'safe distances' no account is taken of magnetic material other than that in the unit under test. If steel parts of a ship's structure are so situated that they form either a magnetic link or a magnetic screen between the compass and the radar set, they may increase or decrease the separation necessary. If any such complication is suspected, a Board surveyor should be consulted.

1.6. If the radar set to be fitted is of a type that has not yet, or has only recently, been type-tested, the compass 'safe distances' may not at the time of fitting be marked on it. The manufacturers of the equipment and the Board's surveyors should be consulted as to the possible effects of the units of the equipment on the compasses and ample allowance should be made for these effects in siting the units.

1.7. Certain radar spares, particularly spare magnetrons, may seriously affect magnetic compasses. Such spares should normally be stored at least 30 feet from magnetic compasses: but if, being spares for a type-tested set, they are housed in a compartment or receptacle that has been designed by the manufacturer of the radar equipment for their storage and is marked with a 'safe distance' allotted by the Admiralty Compass Observatory they may be stored within 30 feet of the compasses but no nearer than the indicated 'safe distance'.

120

2. Effect on other radio installations including D/F

2.1. Radio interference. Whilst direct radio interference from the units of a type-tested radar equipment should not be excessive, it is advisable that all such units, and particularly that containing the modulator, should not be sited within 20 feet of leads from the loop of the direction-finder or from other radio aerials.

2.2. Effect on D/F. Bearing errors may be introduced into a direction-finder if metallic masses, parts of which are higher than the base of the loop, are sited within 6 feet of the loop. In siting radar aerials or radar cabins account should be taken of this important limitation.

3. Siting of aerial unit

3.1. Experience has shown that an aerial height of between 40 and 60 feet offers the best overall radar performance. Increased aerial height will increase the maximum target detection range of the equipment but will at the same time increase the amplitude and extent of 'sea clutter' thus rendering the echoes of buoys and small craft within this area of sea clutter less conspicuous. The aerial should be mounted on a rigid structure which will not twist and give rise to bearing errors.

3.2. Any part of the ship's structure that is at about the same height as the radar aerial may produce a 'shadow sector' on the display, i.e. a sector on the radar screen in which targets may not be seen. The angular width of a shadow sector is determined by the width of the obstruction that causes it and the nearness of that obstruction to the aerial. It is highly desirable so to site the aerial as to avoid shadow sectors or, if this is not possible, to ensure that such shadow sectors as exist will be as narrow as possible and will occur in sectors where they will detract least from the value of the radar as a navigational instrument. Raising the aerial so that it looks over obstructions may be an acceptable measure provided the limitations mentioned in para. 3.1. above are borne in mind. In ships that frequently navigate astern, the need to avoid shadow sectors astern should not be forgotten.

3.3. Reception of signals not directly from the target but by reflection from a part of the ship's structure that causes a shadow sector may result in false echoes appearing on the display within that shadow sector. False echoes may also be caused by parts of the ship's structure which whilst they do not cause shadow sectors, are so disposed as to deflect into the radar aerial energy returning from a target. In each case the false echo will appear to emanate from the relative bearing along which the reflecting object lies. Such false echoes may be eliminated by attaching to the reflecting object a flat metal sheet so placed that its surface reflects radar energy upwards, or a sheet of corrugated metal that will scatter the radar energy, or fitting Radar Absorbent Material (R.A.M.) which will absorb it. These measures will not, of course, remove a shadow sector.

3.4. The needs of servicing in all weathers should be taken into account. If the transmitter is housed immediately below the aerial, some form of platform is desirable so that adjustments may be carried out in situ.

3.5. The aerial unit should be mounted where there is least danger of its being fouled by ropes, bunting, derricks, etc., or of its constituting a hazard to personnel working near it.

4. Siting of the display unit

4.1. Display units may be sited in the wheelhouse, in the chartroom or in both. The following are some of the factors that should be borne in mind in selecting the most convenient site for the unit:

    (i) Magnetic safe distances. The permissible separation of the unit from magnetic compasses may dictate the site.

    (ii) Lighting. The small amount of light issuing from the display unit may be enough to interfere with visual lookout when the wheelhouse is blacked out; and there will be occasions when additional light is needed at the display unit either for comparison of the display with a chart or for running repairs to the unit. Conversely, there will be times when ambient light in the wheelhouse is too strong for effective viewing of the display. Difficulties such as the foregoing may be overcome either by siting the display in the chartroom or by screening it with curtains if it is sited in the wheelhouse.

    (iii) Comparison of radar information with other information. A radar observer will wish to compare his radar display with his charts and with what is seen visually from the wheelhouse. In deciding whether the balance of advantage lies in siting the display in the wheelhouse or the chartroom, the complement of watch-keeping officers available will be an important factor, for it is essential that where only one watchkeeper is on duty he should be able to refer to the radar display and at the same time be immediately available for visual lookout.

    A site in the wheelhouse for the display unit is also convenient for reference by the master or pilot.

    (iv) Viewing facilities. It should be possible for two officers to view a display simultaneously and for one officer to observe for long periods. The height and angle of the display may be such as to make the provision of a seat desirable.

    (v) Direction of view. Navigators have evinced a strong preference for the display unit to be so sited that the observer faces forward when viewing it.

    (vi) It may be advantageous to provide a communication link between the display unit and the transceiver where these units are widely separated.

5. Performance monitor

When a performance monitor carries a permanent echo on the display, the echo should fall in an existing shadow sector or on a bearing of minor importance. Preferably, the performance monitor should be of a type which does not exhibit an echo when it is not activated.

6. Exposed and protected equipment

6.1. Manufacturers of type-tested sets are required to mark each unit of a set as 'Class X' (suitable for fitting in an open space or exposed position) or 'Class B' (for use below deck or in a deckhouse). No units of 'Class B' should be mounted in an open space.

6.2. If a ship is to be fitted with a non-type-tested set, the manufacturer should be asked to state which items have been designed for siting in an exposed position and all units not so designed should be sited below deck or in a deckhouse.

6.3. Care should be taken in siting radar equipment to avoid an environment of excessive heat, fumes or vibration.

7. Lengths of wave-guide and cable runs

When a set is type-tested, the manufacturer is required to provide lengths of cable and wave-guide equal to the maximum lengths that will be employed in a normal installation. If these lengths are exceeded, the performance of the set may be impaired. Manufacturers should therefore be consulted as to the length of wave-guide and cable runs for which a set was designed and the units should be so sited that these lengths are not exceeded.

8. Accident prevention

High Voltage Circuits. Type-tested sets are so designed that there are safeguards that deny ready access to high voltages. Each unit of a non-type-tested set should be so installed as not to constitute a danger either by physical contact or by electric shock to those who handle it.

9. Motor alternators or inverters

9.1. Noise. Though the mechanical noise from units of a type-tested radar equipment will have been reduced to a minimum, care should be taken in the siting of a motor alternator unit to ensure that the noise from it will not interfere with the crew either on or off duty.

9.2. *Heat and/or fumes. Motor alternators or invertors should not be installed in positions where excessive heat, fumes or vibration will cause failure in a relatively short period.*

PRECAUTIONS TO BE TAKEN AFTER FITTING

10. Alignment of heading marker

10.1. A marine radar of a type that has been granted the Board's certificate of type-testing should be capable of measuring the bearing of an object whose echo appears near the edge of the PPI display with an error of no more than 1°. If, however, the commencement of the PPI trace is not correctly centred with the bearing scale or the heading marker is not accurately aligned with the ship's fore and aft line, additional bearing errors will be introduced. It is therefore important that the equipment be correctly set up in these two respects when installed and that its accuracy should be periodically checked. When the heading marker is accurately aligned, a bearing taken by radar should be substantially the same as that obtained visually.

10.2. The following procedures are recommended:

(*a*) Centring the trace.

Each time the radar is switched on, and at the commencement of each watch when the radar is used continuously and whenever bearings are to be measured, the observer should check that the trace is rotating about the centre of the display and should, if necessary, adjust it (*the centre of the display is the centre of rotation of the bearing scale cursor*).

(*b*) Aligning the heading marker and radar aerial.

Visually aligning the radar aerial along what appears to be the ship's fore and aft line is not a sufficiently accurate method of alignment, the following procedure is recommended for accurate alignment:

(i) adjust accurately the centre of rotation of the trace. Switch off azimuth stabilization;
(ii) on equipments possessing the appropriate controls, rotate the PPI picture so that the heading marker lies at 0° on the bearing scale;
(iii) select an object which is conspicuous but small visually and whose echo is small and distinct and lies as nearly as possible at the maximum range of the range scale in use. Measure simultaneously the relative visual bearing of this object and the angle on the PPI that its echo makes with the heading marker; it is important that the visual bearing should be measured from a position near the radar scanner in plan. Repeat these measurements twice at least and calculate the mean difference between bearings obtained visually and by radar;
(iv) if an error exists, adjust the heading marker contacts in the scanner assembly to correct the position of the heading marker by moving it an amount equal to the mean difference calculated in (iii) above;
(v) rotate the PPI picture to return the heading marker to 0° on the bearing scale;
(vi) take simultaneous visual and radar bearings as in (iii) above to check the accuracy of alignment. *Alignment of the heading marker or correcting the alignment on a ship berthed in a dock or harbour, or using bearings of a target that has not been identified with certainty both by radar and visually can introduce serious bearing errors. The procedure for alignment of heading marker should be carried out on clearly identified targets clear of a confusion of target echoes. The alignment should be checked at the earliest opportunity.*

10.3. Checking heading marker alignment. It is recommended that checks should be made periodically in the manner laid down in (*b*) (i), (ii) and (iii) above to ensure that correct alignment is maintained. (Care should be taken to centre the trace accurately beforehand.) If adjustment of the heading marker contacts is required but cannot be carried out immediately, a notice should be displayed prominently calling attention to the existence of an error in heading marker alignment, and in operating the radar due allowance should be made for this error. The alignment should be adjusted at the first opportunity.

11. Measurement of shadow sectors

11.1. The angular width and bearing of any shadow sectors should be determined. For a new vessel, this should be done during trials. In other ships it should be done at the first opportunity after fitting the radar set. When determined, the particulars should be recorded on a tally plate fixed near each display unit of the set.

11.2. Two methods of determining the angular width of a shadow sector are:

(*a*) observation of the behaviour of the echo of a small isolated object, such as a buoy not fitted with a corner reflector or a beacon post, when the ship is turned slowly through 360 degrees at a distance of a mile or so from the object. The display unit should be carefully watched, and the bearings between which the echo from the buoy disappears and reappears taken as indicating the shadow sector or sectors. The sea should be calm so that the echo is not lost in the sea clutter or submerged or hidden by waves from time to time, or in the case of a buoy or other floating object the echo fading temporarily due to any rolling motion.

(*b*) Observation of the shadow sector against a background of sea clutter.

Note.—A shadow sector cannot be fairly estimated in heavy clutter, as echoes from either side of the sector may spread into it and give an illusion that objects in the sector are being observed. Nor can it be satisfactorily determined in confined waters, because of the probability of indirect, false or multiple echoes being produced from nearby buildings or other vessels.

11.3. Calculation of a shadow sector's width and position from a knowledge of the width of the mast or other object causing the shadow and its distance and bearing from the centre of the radar aerial is a useful guide to the shadow sector's expected appearance on the plan display.

11.4. If derricks are found to cause shadow sectors, the effect of their being stowed in more than one position should be observed.

11.5. A change in trim may alter shadow sectors. If visual inspection of the objects causing or likely to cause shadow sectors shows that this is likely, the sectors should be measured under various conditions of trim as opportunity permits.

125

12. D/F apparatus and compasses

The calibration of D/F apparatus on the ship should be checked as soon as practicable after installation of radar equipment; if necessary the apparatus should be re-calibrated.

The accuracy of compass deviation corrections should be checked as soon as practicable after installation of radar equipment even if all units of the equipment are sited outside their prescribed 'safe distances'; if necessary the compass should be adjusted.

If any unit has been sited nearer to a compass than the safe distance, checking the compass should be regarded as essential (*see* paras. 1.2 and 1.3).

13. Mutual interference between radio and radar

Tests should be made with radio receivers and transmitters, working on all frequencies likely to be used, for possible mutual interference between the radar and radio installations.

Further advice on siting precautions in particular installations can be obtained from the Board of Trade, Marine Navigational Aids Branch, or from the marine surveyors at the ports.

Board of Trade
Marine Division
London
April 1968

MNA 9/13/02 Pt 3

APPENDIX B

SPEED, DISTANCE, AND TIMETABLE

For intervals up to 60 minutes

Mins.	\(5\)	\(6\)	\(7\)	\(8\)	\(9\)	Mins.	\(10\)	\(11\)	\(12\)	\(13\)	\(14\)	Mins.	\(15\)	\(16\)	\(17\)	\(18\)	\(19\)	Mins.
1	.1	.1	.1	.1	.2	1	.2	.2	.2	.2	.2	1	.3	.3	.3	.3	.3	1
2	.2	.2	.2	.3	.3	2	.3	.4	.4	.4	.5	2	.5	.5	.6	.6	.6	2
3	.3	.3	.4	.4	.5	3	.5	.6	.6	.7	.7	3	.8	.8	.9	.9	1.0	3
4	.3	.4	.5	.5	.6	4	.7	.7	.8	.9	.9	4	1.0	1.1	1.1	1.2	1.3	4
5	.4	.5	.6	.7	.8	5	.8	.9	1.0	1.1	1.2	5	1.3	1.3	1.4	1.5	1.6	5
6	.5	.6	.7	.8	.9	6	1.0	1.1	1.2	1.3	1.4	6	1.5	1.6	1.7	1.8	1.9	6
7	.6	.7	.8	.9	1.1	7	1.2	1.3	1.4	1.5	1.6	7	1.8	1.9	2.0	2.1	2.2	7
8	.7	.8	.9	1.1	1.2	8	1.3	1.5	1.6	1.7	1.9	8	2.0	2.1	2.3	2.4	2.5	8
9	.8	.9	1.1	1.2	1.4	9	1.5	1.7	1.8	2.0	2.1	9	2.3	2.4	2.6	2.7	2.9	9
10	.8	1.0	1.2	1.3	1.5	10	1.7	1.8	2.0	2.2	2.3	10	2.5	2.7	2.8	3.0	3.2	10
11	.9	1.1	1.3	1.5	1.7	11	1.8	2.0	2.2	2.4	2.6	11	2.8	2.9	3.1	3.3	3.5	11
12	1.0	1.2	1.4	1.6	1.8	12	2.0	2.2	2.4	2.6	2.8	12	3.0	3.2	3.4	3.6	3.8	12
13	1.1	1.3	1.5	1.7	2.0	13	2.2	2.4	2.6	2.8	3.0	13	3.3	3.5	3.7	3.9	4.1	13
14	1.2	1.4	1.6	1.9	2.1	14	2.3	2.6	2.8	3.0	3.3	14	3.5	3.7	4.0	4.2	4.4	14
15	1.3	1.5	1.8	2.0	2.3	15	2.5	2.8	3.0	3.3	3.5	15	3.8	4.0	4.3	4.5	4.8	15
16	1.3	1.6	1.9	2.1	2.4	16	2.7	2.9	3.2	3.5	3.7	16	4.0	4.3	4.5	4.8	5.1	16
17	1.4	1.7	2.0	2.3	2.6	17	2.8	3.1	3.4	3.7	4.0	17	4.3	4.5	4.8	5.1	5.4	17
18	1.5	1.8	2.1	2.4	2.7	18	3.0	3.3	3.6	3.9	4.2	18	4.5	4.8	5.1	5.4	5.7	18
19	1.6	1.9	2.2	2.5	2.9	19	3.2	3.5	3.8	4.1	4.4	19	4.8	5.1	5.4	5.7	6.0	19
20	1.7	2.0	2.3	2.7	3.0	20	3.3	3.7	4.0	4.3	4.7	20	5.0	5.3	5.7	6.0	6.3	20
21	1.8	2.1	2.5	2.8	3.2	21	3.5	3.9	4.2	4.6	4.9	21	5.3	5.6	6.0	6.3	6.7	21
22	1.8	2.2	2.6	2.9	3.3	22	3.7	4.0	4.4	4.8	5.1	22	5.5	5.9	6.2	6.6	7.0	22
23	1.9	2.3	2.7	3.1	3.5	23	3.8	4.2	4.6	5.0	5.4	23	5.8	6.1	6.5	6.9	7.3	23
24	2.0	2.4	2.8	3.2	3.6	24	4.0	4.4	4.8	5.2	5.6	24	6.0	6.4	6.8	7.2	7.6	24
25	2.1	2.5	2.9	3.3	3.8	25	4.2	4.6	5.0	5.4	5.8	25	6.3	6.7	7.1	7.5	7.9	25
26	2.2	2.6	3.0	3.5	3.9	26	4.3	4.8	5.2	5.6	6.1	26	6.5	6.9	7.4	7.8	8.2	26
27	2.3	2.7	3.2	3.6	4.1	27	4.5	5.0	5.4	5.9	6.3	27	6.8	7.2	7.7	8.1	8.6	27
28	2.3	2.8	3.3	3.7	4.2	28	4.7	5.1	5.6	6.1	6.5	28	7.0	7.5	7.9	8.4	8.9	28
29	2.4	2.9	3.4	3.9	4.4	29	4.8	5.3	5.8	6.3	6.8	29	7.3	7.7	8.2	8.7	9.2	29
30	2.5	3.0	3.5	4.0	4.5	30	5.0	5.5	6.0	6.5	7.0	30	7.5	8.0	8.5	9.0	9.5	30
31	2.6	3.1	3.6	4.1	4.7	31	5.2	5.7	6.2	6.7	7.2	31	7.8	8.3	8.8	9.3	9.8	31
32	2.7	3.2	3.7	4.3	4.8	32	5.3	5.9	6.4	6.9	7.5	32	8.0	8.5	9.1	9.6	10.2	32
33	2.8	3.3	3.9	4.4	5.0	33	5.5	6.1	6.6	7.2	7.7	33	8.3	8.8	9.4	9.9	10.5	33
34	2.8	3.4	4.0	4.5	5.1	34	5.7	6.2	6.8	7.4	7.9	34	8.5	9.1	9.6	10.2	10.8	34
35	2.9	3.5	4.1	4.7	5.3	35	5.8	6.4	7.0	7.6	8.2	35	8.8	9.3	9.9	10.5	11.1	35
36	3.0	3.6	4.2	4.8	5.4	36	6.0	6.6	7.2	7.8	8.4	36	9.0	9.6	10.2	10.8	11.4	36
37	3.1	3.7	4.3	4.9	5.6	37	6.2	6.8	7.4	8.0	8.6	37	9.3	9.9	10.5	11.1	11.7	37
38	3.2	3.8	4.4	5.1	5.7	38	6.3	7.0	7.6	8.2	8.9	38	9.5	10.1	10.8	11.4	12.0	38
39	3.3	3.9	4.6	5.2	5.9	39	6.5	7.2	7.8	8.5	9.1	39	9.8	10.4	11.0	11.7	12.4	39
40	3.3	4.0	4.7	5.3	6.0	40	6.7	7.3	8.0	8.7	9.3	40	10.0	10.7	11.3	12.0	12.7	40
41	3.4	4.1	4.8	5.5	6.2	41	6.8	7.5	8.2	8.9	9.6	41	10.3	10.9	11.6	12.3	13.0	41
42	3.5	4.2	4.9	5.6	6.3	42	7.0	7.7	8.4	9.1	9.8	42	10.5	11.2	11.9	12.6	13.3	42
43	3.6	4.3	5.0	5.7	6.5	43	7.2	7.9	8.6	9.3	10.0	43	10.8	11.5	12.2	12.9	13.6	43
44	3.7	4.4	5.1	5.9	6.6	44	7.3	8.1	8.8	9.5	10.3	44	11.0	11.7	12.5	13.2	13.9	44
45	3.8	4.5	5.3	6.0	6.8	45	7.5	8.3	9.0	9.8	10.5	45	11.3	12.0	12.8	13.5	14.3	45
46	3.8	4.6	5.4	6.1	6.9	46	7.7	8.4	9.2	10.0	10.7	46	11.5	12.3	13.0	13.8	14.6	46
47	3.9	4.7	5.5	6.3	7.1	47	7.8	8.6	9.4	10.2	11.0	47	11.8	12.5	13.3	14.1	14.9	47
48	4.0	4.8	5.6	6.4	7.2	48	8.0	8.8	9.6	10.4	11.2	48	12.0	12.8	13.6	14.4	15.2	48
49	4.1	4.9	5.7	6.5	7.4	49	8.2	9.0	9.8	10.6	11.4	49	12.3	13.1	13.9	14.7	15.5	49
50	4.2	5.0	5.8	6.7	7.5	50	8.3	9.2	10.0	10.8	11.7	50	12.5	13.3	14.2	15.0	15.8	50
51	4.3	5.1	6.0	6.8	7.7	51	8.5	9.4	10.2	11.1	11.9	51	12.8	13.6	14.5	15.3	16.2	51
52	4.3	5.2	6.1	6.9	7.8	52	8.7	9.5	10.4	11.3	12.1	52	13.0	13.9	14.7	15.6	16.5	52
53	4.4	5.3	6.2	7.1	8.0	53	8.8	9.7	10.6	11.5	12.4	53	13.3	14.1	15.0	15.9	16.8	53
54	4.5	5.4	6.3	7.2	8.1	54	9.0	9.9	10.8	11.7	12.6	54	13.5	14.4	15.3	16.2	17.1	54
55	4.6	5.5	6.4	7.3	8.3	55	9.2	10.1	11.0	11.9	12.8	55	13.8	14.7	15.6	16.5	17.4	55
56	4.7	5.6	6.5	7.5	8.4	56	9.3	10.3	11.2	12.1	13.1	56	14.0	14.9	15.9	16.8	17.7	56
57	4.8	5.7	6.7	7.6	8.6	57	9.5	10.5	11.4	12.4	13.3	57	14.3	15.2	16.2	17.1	18.1	57
58	4.8	5.8	6.8	7.7	8.7	58	9.7	10.6	11.6	12.6	13.5	58	14.5	15.5	16.4	17.4	18.4	58
59	4.9	5.9	6.9	7.9	8.9	59	9.8	10.8	11.8	12.8	13.8	59	14.8	15.7	16.7	17.7	18.7	59
60	5.0	6.0	7.0	8.0	9.0	60	10.0	11.0	12.0	13.0	14.0	60	15.0	16.0	17.0	18.0	19.0	60

INDEX

Anchoring 109
Anti-clutter 18
Aspect 42
Atmospheric effects 27
Autaplot 115
Avoiding action 72

Beacons 94
Beam width distortion 25
Bearing
    Errors 24
    Steady 36, 73
    Changing 79
Blurring 12
Brilliance 18

Clear weather practice 19, 102
Close quarters 71
Clutter 18
Coastal navigation 85
Collision avoidance 70
Corner reflector 30

Detection range 28
Differentiator 29
Displays 12
Digiplot 116

Early and substantial action 72
Errors
    Bearing 24
    Range 26

False echoes 22

Heading marker 24, 124
Height
    Scanner 28
    Target 28

Ice 29
Indirect echoes 22

Log (Radar) 19

Maladjustment 17
M. Notice 119
Multiple echoes 22

Nearest approach 33
Non-stander propagation 23

Picture quality 16
Performance monitor 17
Plotting
    Relative 41
    True 57
    Questions 59
    Answers 67
Plume 17
Position fixing 88
Pulse length 26

Racon 94
Radar interference 21, 22
Ramark 94
R.A.S. plotter 105
Reflecting properties 26
Reflection plotter 113
Refraction 27
Relative motion 12
Report 45
Routing 83
Rules 9, 32

Scanty information 9
Second trace returns 23
Shadow sectors 21, 125
Side lobe echoes 22
(Siting) 121
Speed 70
Speed and distance table 127
Stopping 85

Target
    Height 28
    Range 26
    Aspect 26
Tide (effects) 92
True plot 57
Turning circle 86
Typical collision case 39

Unseen targets 7

Weather effects 28
Why collisions occur 7